The Role of Theology in the History and Philosophy of Science

Theology

Volumes published in this Brill Research Perspectives title are listed at *brill.com/rpth*

The Role of Theology in the History and Philosophy of Science

By

Joshua M. Moritz

BRILL

LEIDEN | BOSTON

This paperback book edition is simultaneously published as issue 1.2 (2017) of *Theology*, DOI 10.1163/24683493-12340002.

Typeface for the Latin, Greek, and Cyrillic scripts: "Brill". See and download: brill.com/brill-typeface.

ISBN 978-90-04-36021-1 (paperback)
ISBN 978-90-04-36022-8 (e-book)

This book is printed on acid-free paper and produced in a sustainable manner.

Contents

The Role of Theology in the History and Philosophy of Science

Joshua M. Moritz
University of San Francisco, Jesuit School of Theology of Santa Clara University, and the Graduate Theological Union, USA
Joshua.moritz@gmail.com

Abstract

After a bibliographic introduction highlighting various research trends in science and religion, this essay explores how the current academic and conceptual landscape of theology and science has been shaped by the history of science, even as theology has informed the philosophical foundations of science. The first part assesses the historical interactions of science and the Christian faith (looking at the cases of human dissection in the Middle Ages and the Galileo affair) in order to challenge the common notion that science and religion have always been at war. Part two investigates the nature of the interaction between science and Christian theology by exploring the role that metaphysical presuppositions and theological concepts have played—and continue to play—within the scientific process.

Keywords

Science and Religion – Theology and Science – Conflict Thesis – Galileo – Human Dissection – Metaphysical Presuppositions of Science

Bibliographic Introduction to "Science and Religion"

Wayfarers of the 21th century, we are the inhabitants of an increasingly fragmented reality. Customization is the reigning rule of a culture dominated by internet individualism, as playlists replace albums, Netflix supplants the major networks, and smartphones supersede neighbors. As the deconstruction of metanarratives continues in academia and the numerous sub-disciplines of

 | DOI 10.1163/9789004360228

the natural sciences become more and more specialized every day, the tides of postmodernity continue to rise, creating countless islands of isolated individuals. Postmodern deconstruction, critical consciousness, and the hermeneutics of suspicion have assailed the bastions of Enlightenment rationality and have shaken the social foundations of scientific authority.[1] While many feel that these forces have "left us bereft and alone amid a desert of meaninglessness,"[2] others have sought refuge in the metanarratives promised by the evangelists of pseudoscientific scientism and the showmen of post-scientific young earth creationism.[3] In the wake of such multifarious forces of fragmentation, the field of theology and science endeavors to reintegrate meaning, rediscover truth, and begin to put the pieces back together through a *post-critical reconstruction* of reality. Even as scientists have been shown by philosophers that science *is not* (and *should not* be) value-free, theologians have increasingly come to realize that their theological constructions *cannot* (and *should not*) be divorced from the scientific models that describe physical reality.[4] To speak intelligibly of God as either timeless or temporal requires that one have a realistic understanding of physical time. To preach that God "so loved the cosmos ..." requires an informed conception of the scientific character of the cosmos. To speak of the Word becoming flesh demands a deep enquiry into the nature of biological life. And so on.

The academic field of 'science and religion' (also known as 'theology and science') has a much wider scope than most interdisciplinary fields. Because the central questions within the natural sciences are constantly changing, the foci of theology and science are always shifting as well. One of the best ways to get an overall sense of where the field has been and where it is going is to look at the topics and speakers of key distinguished lectureships such as the Gifford Lectures, the Boyle Lectures, the Goshen Lectures in Science and Religion, and

1 For a discussion of the postmodernist critique of science and a number of responses see *A House Built on Sand: Exposing Postmodernist Myths about Science*, ed. Noretta Koertge (New York: Oxford University Press, 1998).

2 Ted Peters, *God–The World's Future: Systematic Theology for a Postmodern Era* (Minneapolis: Fortress Press, 1992), 30.

3 For discussion of postmodernism and scientism see Robert Eaglestone, "Postmodernism and ethics against the metaphysics of comprehension" in *The Cambridge Companion to Postmodernism*, ed. Steven Connor (Cambridge: Cambridge University Press, 2004), 182–195, 191; For discussion of postmodernism and creationism see Robert T. Pennock, "The Postmodern Sin of Intelligent Design Creationism," *Science & Education* 19:6 (June 2010), 757–778.

4 See John C. Polkinghorne, *One World: The Interaction of Science and Theology* (Philadelphia: Templeton Foundation Press, 2007).

the J.K. Russell Fellowship. Browsing through these lectures one sees a timeline of the relevant issues at the interface of science and religion. In the first decade of the 20th century one finds American psychologist William James endeavoring to scientifically understand mysticism and other types of religious phenomena in *The Varieties of Religious Experience*.[5] In the 1920's *The Nature of the Physical World*, by physicist and devout Quaker, Arthur Eddington, discusses the philosophical and religious implications of fundamental changes in scientific thought related to areas such as the theory of relativity, quantum theory and thermodynamics.[6] Later in the decade, mathematician and philosopher Alfred North Whitehead's *Process and Reality* develops a philosophy of organisms that places consciousness at the core of reality.[7] Catholic philosopher and historian Etienne Gilson and Reformed theologian Karl Barth dominate the theological scene of Gifford Lectures of the 1930s.[8] And in the same decade physician and theologian Albert Schweitzer aims to find a basis for a common ethics and spirituality in the experiences of those outside of the Christian tradition (*The Problem of Natural Theology and Natural Ethics*).[9] The questions in the science and religion dialogue of the 1940s are overshadowed by the implications of atomic physics, and thus Niels Bohr's lecture *Causality and Complementarity: Epistemological Lessons of Studies in Atomic Physics* is representative of this age.[10] The 1950's witnesses the dramatic social ascendency of the natural sciences which is in turn met by theologians such as Rudolf Bultmann (*History and Eschatology: The Presence of Eternity*) and Paul Tillich (*Systematic Theology*) who seek to insulate theology from the theories of science by focusing on demythologizing the Christian scriptures and discerning the symbolic value of Biblical texts.[11] At the same time Catholic chemist and philosopher of science Michael Polanyi explores the role of subjectivity and

5 William James, *The Varieties of Religious Experience: A Study in Human Nature* (New York: Modern Library, 1902).

6 Arthur Eddington, *The Nature of the Physical World* (New York, Macmillan, 1928).

7 Alfred North Whitehead, *Process and Reality: An Essay in Cosmology* (New York: Free Press, 1929).

8 Karl Barth, *The Knowledge of God and the Service of God according to the Teaching of the Reformation Recalling the Scottish Confession of 1560* (Aberdeen: University of Aberdeen, 1937–1938).

9 Albert Schweitzer, *The Problem of Natural Theology and Natural Ethics* (unpublished).

10 Niels Bohr, *The Philosophical Writings of Niels Bohr, Vol. 4: Causality and Complementarity: Epistemological Lessons of Studies in Atomic Physics The 1949–50 Gifford Lectures* (Woodbridge, CT: Ox Bow Press, 1999).

11 Rudolf Bultmann, *The Presence of Eternity: History and Eschatology* (New York: Harper, 1957); Paul Tillich, *Systematic Theology* (Chicago: University of Chicago Press, 1951–1963).

values within the objective sciences in his *Personal Knowledge: Towards a Post-Critical Philosophy*.[12] While there is a gap in the science and religion focused Gifford lectures of the 1960's, the latter half of this decade is clearly dominated by physicist, philosopher, and theologian Ian G. Barbour's groundbreaking book *Issues in Science and Religion* (1966), which explores how the history and philosophy of science provides a methodological bridge via which theology and science can communicate.[13] Barbour's *Issues* develops an intellectual basis for the facilitating and consolidating of a sustained and systematic dialogue between science and religion and, in this way, this book has been widely credited with creating the contemporary field of science and religion.

The 1970s witnessed a broadening of Barbour's work on history and philosophy as the methodological foundations of the science and religion dialogue continued to be developed on a number of fronts. The Gifford Lectures of historian of science Reijer Hooykaas (*Fact, Faith and Fiction in the Development of Science*) views 'faith' as something larger than religion. For him faith refers to broad *a priori* conceptions that are operative at any given moment.[14] According to Hooykaas, faith is what motivates a scientist to investigate in the first place, and faith informs the framework of knowledge within which the investigation takes place. In a similar way, 'facts' are not seen as simply objective and immutable truths, but rather, facts are phenomena which exist as something external to the individual subject. Meanings (or facts) can be extracted from these external phenomena, but such meanings must always be interpreted in light of faith. According to Hookyaas, 'fictions' are hermeneutical intellectual tools or theories that are judged as true or false through the scientific process. In *The Road of Science and the Ways to God* physicist and historian of science Stanley Jaki aims to demonstrate how the emergence of science in the West necessary depended upon the metaphysical and cultural matrix provided by the Judeo-Christian worldview.[15] Such worldview requirements for the rise of science include a conviction that the universe is rational, intelligible, and contingent. These faith affirmations, according to Jaki, serve as permanent and unchangeable metaphysical foundations for science, and they must be present

12 Michael Polanyi, *Personal Knowledge: Towards a Post-Critical Philosophy* (Chicago, University of Chicago Press, 1958).

13 Ian G. Barbour, *Issues in Science and Religion* (Englewood Cliffs, N.J., Prentice-Hall, 1966).

14 Reijer Hooykaas, *Fact, Faith and Fiction in the Development of Science: The Gifford Lectures Given in the University of St. Andrews 1976* (Dordrecht; Boston: Kluwer Academic Publishers, 1999).

15 Stanley Jaki, *The Road of Science and the Ways to God* (Chicago: University of Chicago Press, 1978)

as a necessary minimum for the existence and continued success of science. At the close of the decade neuroscientist and Nobel Laureate Sir John Eccles (*The Human Mystery* and *The Human Psyche*) critiques scientific reductionism as an inadequate philosophical framework for understanding the mystery of the human mind and Eccles is skeptical that materialism will eventually be able to account for the entirety of spiritual reality and experience in terms of patterns of neuronal activity.[16] Instead of viewing the mystery of the human person as something that can be reduced to its physical constituents, measured by the tools of science, and ultimately quantified, Eccles argues for a mind-body dualism (*dualist-interactionism*) where a nonmaterial mind or self is thought to be both *acting upon* and *influenced* by a material brain.

In the 1980s philosopher Richard Swinburne continues to explore the nature of the human self in *The Evolution of the Soul* and, like Eccles, defends a dualistic perspective that sees mental states as distinct from brain states even though such mental states both *cause* and are *caused by* brain states.[17] In *Behind the Eye* neuroscientist, physicist, and information theorist Donald M. MacKay desires to offer an alternative to mind-body dualism and focus instead on how mental descriptions and physical descriptions can be seen as *complementary*.[18] In the same way that mathematical properties cannot be reduced to the physical properties of electricity, for MacKay, mental properties cannot be entirely reduced to the physical properties of the brain. While MacKay makes a distinction between different levels of causation he also affirms that properties of events at the mental level of description are necessarily connected to properties of the same event at the physical levels of description. As MacKay strives to move beyond dualism in his understanding of human personhood and identity, Reformed theologian Jürgen Moltmann seeks to move beyond dualistic understandings of the Christian doctrine of creation in his Gifford lectures, *God in Creation: A New Theology of Creation and the Spirit of God*.[19] Moltmann develops what he terms a *pan-en-theistic* understanding of God and cosmos, where God's life-giving Spirit indwells all of creation and all created beings. In ceasing to absolutely distinguish between God and the creation, Moltmann aims to challenge those suffering from scientific disenchantment who see the world as a godless mass of resources that finds worth merely in its potential

16 John Eccles, *The Human Mystery* (Berlin: Springer. 1980); and John Eccles, *The Human Psyche* (New York: Routledge, 1992).

17 Richard Swinburne, *The Evolution of the Soul* (Oxford: Oxford University Press, 1986).

18 Donald M. MacKay, *Behind the Eye* (Oxford: Blackwell, 1991).

19 Jürgen Moltmann, *God in Creation: A New Theology of Creation and the Spirit of God*, 1st Fortress Press ed., The Gifford Lectures 1984–1985 (Minneapolis: Fortress Press, 1993).

for technological exploitation by human beings. Another theologian who dominates the science and religion scene in the 1980s is Wolfhart Pannenberg. Pannenberg argues that theology must not be separated from the natural sciences in the way that Barth, Bultmann, and Tillich argue, but rather theology must actively engage with the sciences and even be accountable to scientific theories and findings.[20] Credible talk about God, says Pannenberg, has to be related to the very reality that is claimed to be God's creation. In the mid-1980's the publication of *God and Nature: Historical Essays on the Encounter between Christianity and Science*—an anthology from a conference at the University of Wisconsin-Madison in 1981—was a watershed work showing that the complexity of historical interactions between science and religion depart from an interpretation of history that sees their interaction in terms of mutual hostility and conflict.[21]

In the 1990's Oxford historian of science John Hedley Brooke further develops the thesis of *God and Nature* in his now classic book, *Science and Religion: Some Historical Perspectives*.[22] Brooke, together with historian of science Geoffrey Cantor, delivered the 1995–6 Gifford Lectures (*Reconstructing Nature*) where they examine a variety of ways in which the relationship between science and religion has been presented throughout history.[23] Contending that neither science nor religion is reducible to some timeless "essence," Brooke and Cantor show how science and religion have shaped each other in the past and suggest ways how they might continue to form and inform each other in the future. Molecular biologist Arthur Peacocke (whose Gifford lectures were published as *Theology for a Scientific Age*) agrees with Brooke and Cantor and argues that if theology is to remain grounded in reality, it must seek to find ways to embrace and incorporate the concepts and knowledge derived from the natural sciences.[24] For Peacocke, biological evolution is God's way of creating life, and the ongoing evolutionary process is nothing less than the manifestation of God's continuing interaction with creation. To facilitate the dialogue

---

20 Wolfhart Pannenberg, *Toward a Theology of Nature: Essays on Science and Faith*, trans. Ted Peters (Nashville: Westminster/John Knox Press, 1993).

21 David Lindberg and Ronald Numbers, eds., *God and Nature: Historical Essays on the Encounter between Christianity and Science* (Berkeley: University of California Press, 1986).

22 John Hedley Brooke, *Science and Religion: Some Historical Perspectives* (Cambridge: Cambridge University Press, 1991).

23 John Hedley Brooke and Geoffrey Cantor, *Reconstructing Nature: The Engagement of Science and Religion* (Edinburgh: T & T Clark, 1998).

24 Arthur Peacocke, *Theology for a Scientific Age: Being and Becoming—Natural, Divine, and Human* (Minneapolis: Fortress, 1993).

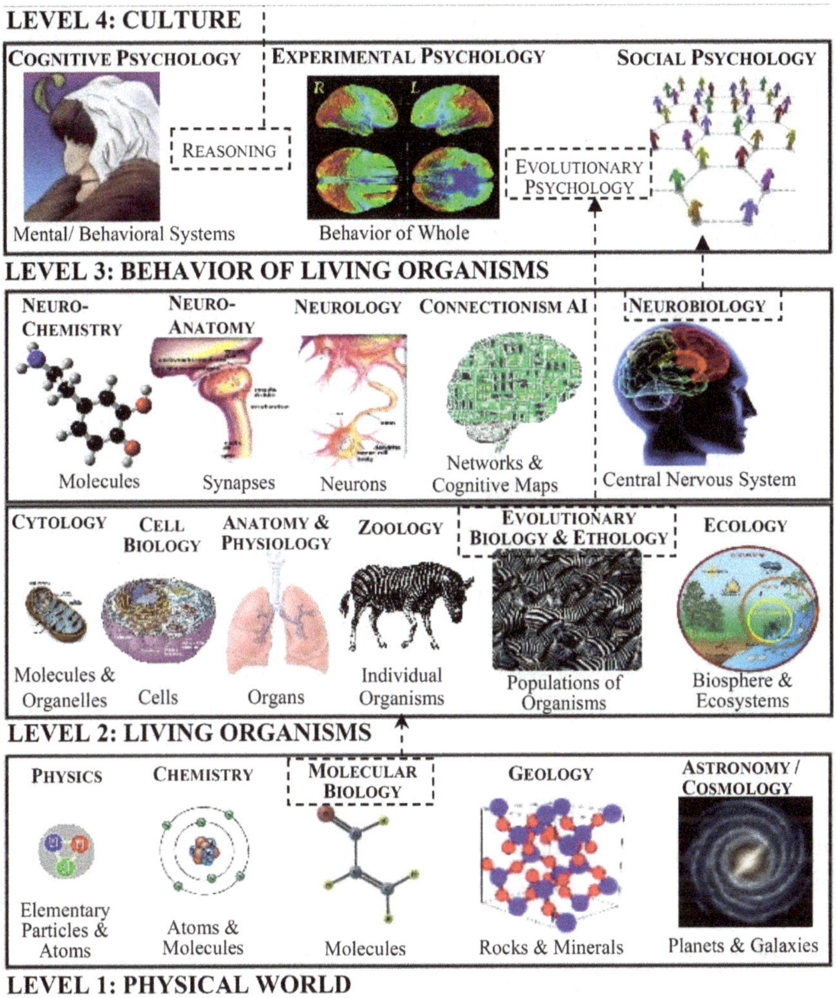

FIGURE 1 *The non-reducible hierarchy of the sciences: The vertical axis lists physical, biological, neurological, behavioral, and cultural phenomena in order of increasing complexity, ranging from atoms to human behavior, and the disciplines which study them. The horizontal axis lists the phenomena being studied at each different level. For levels 1–3 the phenomena being studied within each discipline are listed in order of increasing size from left to right.*

between theology and science Peacocke introduces an *epistemic hierarchy* (see figure) where the bottom levels of the hierarchy (atomic physics, chemistry, molecular biology, etc.) place epistemic constraints on the higher levels (ecology, neurobiology, cognitive psychology, etc.), making the knowledge or ideas from the upper levels philosophically responsible or accountable to those from the lower levels. According to Peacocke's model, each distinct level maintains its own epistemic integrity and the knowledge claimed by a higher level reality cannot be explained by, or reduced to, a lower-level reality. Each of the hierarchical levels of complexity is real and each must be interpreted and explained by the methods and concepts appropriate to it. Peacocke's hierarchical epistemology thus leads to a *philosophy of emergence* that rejects both mind-body dualism and materialistic reductionism.

The 1990s was also the golden age of the genetic research, and thus many discussions in religion and science during this period focus on the theological and religious implications of discoveries in genetics. Philosopher and theologian Holmes Rolston III (*Genes, Genesis and God*), theologian Ted Peters (*Playing God?: Genetic Determinism and Human Freedom*) and geneticist R.J. Berry (*Gods, Genes, Greens and Everything*) are prominent scholars of this decade who argue against genetic determinism.[25] Arguing that cumulative transmissible cultures are unique to *Homo sapiens*, who alone have the capacity for language, Rolston explores how gene-culture co-evolution gives rise to phenomena that once they emerge cannot be reduced to genes alone. Ethics, argues Rolston, is distinctively a product of the human culture and thus Darwinian "evolutionary ethics" and genetic explanations for moral behavior are inadequate. Like ethics, Rolston argues that religious behavior is culturally acquired and not genetically innate.

While discussions about genetics occupied those who were seeking dialogue between biology and religion, others at this time were exploring the implications of quantum physics in order to better understand the nature of causality and produce new models of how God may act in the world. Taking up such questions, scientists and theologians led by physicist and theologian Robert John Russell at the Center for Theology and the Natural Sciences (CTNS) collaborated with scholars at the Vatican Observatory to initiate a decade long research project that would lead to five volumes filled with chapters by leading scholars in science and religion.[26] In the mid-1990s physicist and

25 Holmes Rolston III, *Genes, Genesis and God: Values and their Origins in Natural and Human History* (Cambridge UK: Cambridge UP, 1999); Ted Peters, *Playing God?: Genetic Determinism and Human Freedom*, 2nd ed. (New York and London: Routledge, 2003).

26 *Chaos and Complexity: Scientific Perspectives on Divine Action*, ed. Robert J. Russell, Nancey C. Murphy and Arthur R. Peacocke (Vatican City State; Berkeley, CA: Vatican Observatory Publications; Center for Theology and the Natural Sciences, 1995); *Evolutionary Molecular Biology: Scientific Perspectives on Divine Action*, ed. Robert John Russell, Francisco J. Ayala, William R. Stoeger, SJ, (Vatican City State: Vatican Observatory Publications / Berkeley, CA: CTNS, 1998); *Scientific Perspectives on Divine Action: Quantum Cosmology and the Laws of Nature*, ed. Robert J. Russell, Nancey Murphy, and C.J. Isham

theologian John Polkinghorne presents a 'state of the discipline' address in his *Science and Christian Belief*, where, via a scientific exploration of the substance of the Christian faith, he shows how orthodox Christian belief is compatible with knowledge gained from the natural sciences.[27] At the close of the decade physicist Russell Stannard (*The God Experiment*) seeks to show the compatibility of theology and science by providing a scientifically coherent framework of evidence for the existence of God.[28]

Science and religion scholars of the first decade of the second millennium continue to focus on human uniqueness as theologian J. Wentzel van Huyssteen's *Alone in the World?* argues that scientific understandings of human uniqueness can help to ground theological notions of human distinctiveness such as the doctrine of the *imago Dei*.[29] Van Huyssteen is particularly interested in paleoanthropological evidence for a behavioral revolution in the Upper Paleolithic period when, he says, we find the origins of the embodied human mind as evidenced in spectacular cave paintings that reveal a complex cognition underlying symbolic expression. In *Mind and Emergence* philosopher and theologian Philip Clayton explores how a philosophy of emergence can inform theological anthropology.[30] Philosopher and theologian Nancey Murphy likewise employs philosophical emergence to make a case for an understanding of the soul that is both non-reductive and physical (*nonreductive physicalism*).[31] Theologian John Haught explores the theological implications of God creating

---

(Berkeley, CA: CTNS and Vatican Observatory Publications, 1999); *Neuroscience and the Person: Scientific Perspectives on Divine Action*, ed. Robert John Russell, Nancey Murphy, Theo Meyering, and Michael Arbib (Berkeley, CA: CTNS / Vatican City State: Vatican Observatory, 1999); Quantum Mechanics: Scientific Perspectives on Divine Action, ed. Robert J. Russell, Philip Clayton, Kirk Wegter-McNelly, and John Polkinghorne (Vatican City State; Berkeley, CA: Vatican Observatory Publications; Center for Theology and the Natural Sciences, 2001).

27 John Polkinghorne, *The Faith of a Physicist: Reflections of a Bottom-Up Thinker, The Gifford Lectures, 1993–4* (Princeton, NJ: Princeton University Press, 1994).

28 Russell Stannard, *The God Experiment: Can Science Prove the Existence of God?* (Paulist Press, 2016).

29 J. Wentzel van Huyssteen, *Alone in the World?: Human Uniqueness in Science and Theology* (Grand Rapids: Eerdmans, 2006).

30 Philip Clayton, *Mind and Emergence: From Quantum to Consciousness* (New York: Oxford UP, 2004).

31 Warren S. Brown, Nancey Murphy and H. Newton Maloney, eds. *Whatever Happened to the Soul?: Scientific and Theological Portraits of Human Nature* (Minneapolis: Fortress, 1998); Nancey Murphy, *Bodies and Souls or Spirited Bodies?* (Cambridge: Cambridge University Press, 2006); Nancey Murphy, George Ellis, Timothy O'Connor, *Downward Causation and the Neurobiology of Free Will* (Springer, 2009).

through evolution in *God After Darwin: A Theology of Evolution*.[32] Haught takes Darwinian evolution through natural selection into the very center of his theological reflections on the meaning of life, of God, and of the universe, and finds that while Darwinian evolution offers exciting new solutions to old theological questions it also raises new theological problems. One of the theological dilemmas raised by the idea of God creating through evolution is how pain, suffering, and extinction are intrinsic to the evolutionary process. This issue is discussed in a number of books during this period, including theologian Christopher Southgate's *The Groaning of Creation*, philosopher Michael Ruse's *Can a Darwinian Be a Christian?* and the CTNS anthology *The Evolution of Evil* which includes a wide range of chapters by leading scholars in the field (including chapters by both Southgate and Ruse).[33] As some scholars in theology and science were coming to terms with how Darwinian evolution unearths a deeper dimension to the problem of evil, others such as biochemist and theologian Alister McGrath (*A Fine-Tuned Universe*) and physicist John Barrow (*Fitness of the Cosmos for Life*) were exploring how the physics of the early universe appears to have been fine-tuned for the evolutionary emergence of intelligent life.[34] Murphy's edited volume *Physics and Cosmology: Scientific Perspectives on the Problem of Natural Evil* brings the insights of cosmic fine-tuning to bear on the problem of natural evil.[35]

In the second decade of the 21st century researchers in the Cognitive Science of Religion (CSR) such as Justin Barrett have argued that certain aspects of religious thought are cognitively intuitive (*Born Believers*).[36] Scientists and theologians are beginning to explore non-Darwinian themes in evolutionary biology such as the role of cooperation in evolution (Sarah Coakley, *Sacrifice Regained: Evolution, Cooperation and God*) and the phenomenon of

32 John Haught, *God After Darwin: A Theology of Evolution* (Boulder, CO: Westview Press, 2000).

33 Christopher Southgate, *The Groaning of Creation: God, Evolution, and the Problem of Evil* (Louisville, KY: Westminster John Knox, 2008); Michael Ruse, *Can a Darwinian Be a Christian? The Relationship Between Science and Religion* (Cambridge: Cambridge University Press, 2001); *The Evolution of Evil*, ed. Gaymon Bennett, Martinez J. Hewlett, Ted Peters, and Robert John Russell (Gottingen: Vandenhoeck & Ruprecht, 2008).

34 Alister E. McGrath, *A Fine-Tuned Universe: The Quest for God in Science and Theology* (Louisville, KY: Westminster John Knox, 2009); *Fitness of the Cosmos for Life: Biochemistry and Fine-Tuning*, ed. John D. Barrow (Cambridge: Cambridge University Press, 2008).

35 *Physics and Cosmology: Scientific Perspectives on the Problem of Natural Evil*, ed. Nancey Murphy, Robert J. Russell, and William Stoeger (Vatican City State: Vatican Observatory Publications, 2007).

36 Justin L. Barrett, *Born Believers, The Science of Children's Religious Belief* (New York: Free Press, 2012).

evolutionary convergence (Simon Conway Morris, *Life's Solution* and *The Runes of Evolution*).[37] Numerous scholars (e.g. Celia Deane-Drummond and Joshua Moritz) are making a case that humans are not as evolutionarily unique as previous generations of scholars have thought.[38] And some scholars, such as Ted Peters and David Wilkinson, have turned their gaze to the stars to investigate how the emerging fields of astrobiology, the discovery of habitable extrasolar planets, and the prospect of finding extraterrestrial life changes our theological understanding of the place of humans in the cosmos.[39]

This brief survey of the field of Science and Religion shows how in the fifty years since the publication of Barbour's *Issues* (1966) the discipline has been expanding in scope and growing in the degree to which it has been able to network and connect diverse areas of specialization in a great number of research sub-disciplines. The field of science and religion now lays claim to a number of peer-reviewed journals (e.g. *Zygon: A Journal of Religion and Science, Theology and Science, Science and Christian Belief,* and *Philosophy, Theology and the Sciences*), endowed professorships (e.g. Oxford University, Princeton Theological Seminary, the Graduate Theological Union, and Davidson College), and discipline-specific encyclopedias and companions (e.g. Oxford, Cambridge, Blackwell, Routledge, MacMillan).

Still, in the minds of many academics, the field of Science and Religion seems novel, and scholars often question the legitimacy of novel interdisciplinary ventures. While many new interdisciplinary fields are initially hailed with skepticism, though, doubts within the academy typically subside within a few decades. For example, the field of Business Ethics was initially hailed as "oxmoronic," "without credentials," and "without conceptual subject matter of its own" but now is standard academic fare in philosophy and ethics departments

---

37 Sarah Coakley, *Sacrifice Regained: Evolution, Cooperation and God* (Oxford: Oxford University Press, 2017); Simon Conway Morris, *Life's Solution: Inevitable Humans in a Lonely Universe* (Cambridge: Cambridge University Press, 2003) and Simon Conway Morris, *The Runes of Evolution: How the Universe Became Self-Aware* (West Conshohocken, Templeton Press, 2015).

38 See Celia Deane-Drummond's Goshen Lectures, *Re-imaging the Divine Image: Humans and Other Animals* (Kitchener, Ontario: Pandora Press, 2014) and Joshua Moritz's Goshen Lectures, *Human Uniqueness and the Image of God: Bridging Biology and Theology through the Hebrew Concept of Election* (Kitchener, Ontario: Pandora Press, 2018).

39 David Wilkinson, *Science, Religion, and the Search for Extraterrestrial Intelligence* (Oxford: Oxford University Press, 2013); *Astrotheology: Science and Theology Meet Extraterrestrial Life*, ed. Ted Peters, Robert John Russell, and Joshua Moritz (Eugene, Oregon: Wipf and Stock, 2017).

around the world.[40] In a similar manner, while the field of Sociobiology was quite controversial when it began to take shape as a new area of research the mid-1970s, "Sociobiology is now part of the core research and curriculum of virtually all biology departments, and it is a foundation of the work of almost all field biologists."[41] The vast majority of interdisciplinary areas of research eventually reach a non-controversial status, and come to be accepted within the academy without much resistance or emotional reaction.

But what about the field of science and religion? Although its academic beginnings date back to over a century ago, the scholarly pursuit of relating science and religion is still frequently greeted with marked suspicion and even contempt.[42] The reason for these misgivings is primarily related to persistence of misinformation about the history of science and religion, and continued confusion regarding the central role that theology has played (and still plays) within the philosophical foundations of the natural sciences. To truly understand the nature of the dialogue between theology and science, one must first understand something of how the history of science and religion and how philosophy within science have shaped the current academic and conceptual landscape. The remainder of this essay will thus examine the role of religion and theology within the history and philosophy of the natural sciences, focusing mainly on the Christian faith tradition. The first part will assess the historical interaction of science and the Christian religion in order to challenge the common notion that science and religion (or theology and science) have always been at war. Part two will then investigate the nature of the interaction between science and Christian theology by exploring the role that metaphysical presuppositions and theological concepts have played—and continue to play—within the scientific process.

Part 1: Theology and the Historiography of Science

Our conception of the past significantly shapes our understanding of the present and influences the course of our future. Where we think we came from directly impacts who we are and where we believe we can go. The academic

40 Robert C. Solomon "Business Ethics" in *A Companion to Ethics*, ed. Peter Singer (Malden, Mass.: Blackwell, 1991), 354.

41 Edward H. Hagen, "Controversial Issues in Evolutionary Psychology," in *The Handbook of Evolutionary Psychology*, edited by David M. Buss (Hoboken, NJ: Wiley, 2005), 167.

42 For example The Gifford Lectures are one of most prestigious honors in Scottish academia. These lectures, which focus on the relationship between religion and science, were founded in 1898 and are awarded annually.

discipline of Science and Religion (aka Theology and Science) is an interdisciplinary field of research with a deep and complex history that implicitly imposes a framework that structures and informs current discussions. Yet, few among non-specialists are aware of the subtleties and details of this history. A majority of the public believes that the history of science and religion has been primarily characterized by conflict.[43] This notion of an essential conflict (known as the "conflict thesis" or "warfare thesis") has been—and continues to be—reinforced through entertainment, media presentations, and the popular writings of anti-religious scientists to a point where many people feel that nothing could be more obvious than the reality of a continuous battle between science and religion. Even though professional historians of science and religion have insisted for decades that there never was such a war in the past, their scholarly research and writings have had little impact on the general feeling that science and religion have always been in a state of perpetual conflict. Beyond the popular dimension the conflict thesis "persists in various academic disciplines, and it is still presented to university students as the authoritative historical account of religion–science interactions."[44] As a consequence, the warfare thesis "continues to exercise a tenacious hold on the popular imagination and still informs many nonspecialist accounts of science and its history."[45]

The commonly held notion that science and religion are somehow opposites, or that theology and science are—and have always been—at war, poses a difficult challenge to scholars in the field because this perception acts as a significant barrier when scholars in science and religion interact with researchers in other disciplines or when they communicate the findings of their field with the general public. Because there is a popular perception that science and religion are always at odds, or that they have nothing meaningful to say to one another, scholars in the field are often forced to take an apologetic stance when discussing their research with those outside the discipline. Understanding the relationship between science and religion requires investigating the historical foundations of their interaction and recognizing how this history challenges common presuppositions that have emerged from the popular discourse asserting their continual warfare. To help lay the foundations for dialogue this

43 For example, a 2015 Pew Research Foundation poll found that 73% of adults who seldom or never attend religious services believe that science and religion are in conflict. Cary Funk and Becka A. Alper, "Religion and Science," 2015 Pew Research Foundation Poll, http://www.pewinternet.org/2015/10/22/science-and-religion/ (October 22, 2015).

44 Thomas Aechtner, "Galileo Still Goes to Jail: Conflict Model Persistence within Introductory Anthropology Materials," *Zygon*, 50:1 (2015), 209–226, 209.

45 Peter Harrison, *The Territories of Science and Religion* (Chicago: University of Chicago Press, 2014), 24.

chapter will examine the roots of the conflict thesis to show that this thesis has little—if any—basis in historical fact. The real story of the historical interaction of "science" and "religion" is complex. The way in which we understand the terms "science" and "religion" is a relatively modern phenomenon and our current usage of the terms "science" and "religion" is itself the product of an ideological movement which began in the 19th century with the explicit intention of secularizing science and privatizing religion.[46] Yet, even if we anachronistically project our current definitions back into history one finds that where tensions *can* be found, such tensions are typically not *between* science and religion (or theology) *as such*, but rather between what may be called a conservative or cautious intellectual attitude and an innovative or speculative intellectual attitude.[47] And even this type of tension is more often than not, a *creative* tension that has worked to refine empirical methods and further scientific knowledge.

The Historical Roots of the Conflict Thesis

The historical roots of the conflict thesis lie in mid-19th century America and Britain.[48] According to historian of science Ronald Numbers, "The historiographical war between science and religion began in 1869," when the founding president of Cornell University, Andrew Dickson White, "lectured to a large audience at the Cooper Union in New York City on 'The Battle-Fields of Science.'"[49] In the succeeding decades White expanded this lecture first into a short book, *The Warfare of Science* (1876), and eventually into a two volume tome, *A History of the Warfare of Science with Theology in Christendom* (1896). White himself was the advance guard of a movement that believed that the professionalization and progress of the natural sciences required the secularization of science. In his earlier days White had come to espouse the historicist vision of the

46 As John Hedley Brooke argues, "There is no such thing as *the* relationship between science and religion." *Science and Religion: Some Historical Perspectives* (New York: Cambridge University Press, 1991), 438.

47 Maurice A. Finnochario, "Science, Religion, and the Historiography of the Galileo Affair: On the Undesirability of Oversimplication" *Osiris, Vol. 16, Science in Theistic Contexts: Cognitive Dimensions* (2001), 114–132, 116.

48 As Peter Harrison comments "It is now generally accepted by historians that this erroneous view, known as 'the conflict myth', was largely the invention of two nineteenth-century controversialists, John Draper and Andrew Dickson White." Peter Harrison, "Introduction," *Cambridge Companion to Science and Religion*, ed. Peter Harrison (Cambridge: Cambridge University Press, 2010), 4.

49 Ronald Numbers, "Science and Religion," *Osiris, Vol. 1, Historical Writing on American Science* (1985), 59–80, 59.

evolution of culture propounded by the positivist philosopher Auguste Comte where theology represented the most primitive form of cultural development and the natural sciences stood as the telos and endpoint of civilized society.[50] Convinced that theologically affiliated education impeded the development of the type of progressive and evolutionary view of science that he favored, White thus sought to establish Cornell College as a non-sectarian university that would serve as the paradigm of secular and positivist scientific education. White's vision of a purely secular university "stood in sharp contrast to the traditional religious sponsorship of colleges and universities."[51] Receiving severe criticism from religious figures while competing with religiously affiliated colleges for government funding—White needed to demonstrate why his own university model was better than that of his religious rivals.[52] Thus, White asserted the thesis in his 1896 best-seller that "In all modern history, interference with science in the supposed interest of religion, no matter how conscientious such interference may have been, has resulted in the direst evils both to religion and to science."[53]

White's conflict thesis was passionately echoed and emotionally elaborated upon by English-born, New York City chemist, John William Draper in his *History of the Conflict between Religion and Science* (1874). Asked to contribute a volume to a popular International Scientific Series published by Yeomans Press, Draper produced a graphic and dramatic narration depicting the relationship between science and Christian theology as a timeless and violent struggle between good and evil.[54] "Science," says Draper, "has never subjected any one to mental torment, physical torture, least of all to death, for the purpose

---

50 Peter Harrison, *Territories*, 174.

51 David B. Wilson, "The Historiography of Science and Religion," in *The History of Science and Religion in the Western Tradition*, ed. Gary B. Ferngren (New York: Garland, 2000) 3.

52 White later reflected, "Nearly all the denominational colleges girded themselves for the fray, and sent their agents to fight us at Albany ... For a long time I stood on the defensive, but as ... this seemed only to embitter our adversaries, I finally determined to take the offensive." James R. Moore, *The Post-Darwinian Controversies: A Study of the Protestant Struggle to Come to Terms with Darwin in Great Britain and America, 1870–1900*, (Cambridge: Cambridge University Press, 1979), 34.

53 Andrew Dickson White, *A History of the Warfare of Science with Theology in Christendom*, reprint edition (Gloucester, MA: Peter Smith, 1978), viii.

54 At this time, Edward Livingston Youmans was "America's chief purveyor of popular science." Moore, *The Post-Darwinian Controversies*, 20; No other title in the hundred-odd volume International Scientific Series sold as well. In the United States *The Conflict* passed through fifty printings in about as many years. In Great Britain there were twenty-one editions in fifteen years. Elsewhere it achieved comparable notoriety in French, German,

of upholding or promoting her ideas," and is "unstained by cruelties and crimes." Religion, however, is "steeped in blood!"[55] For Draper, though, not all religions are equally culpable. While the Roman Catholic Church is seen as the chief historical antagonist to science, and Protestants are portrayed as a slightly lesser evil, Draper champions Islam as a fearless, freethinking, empirically-minded advocate of science.[56] Draper tells his readers that while the Christian West descended into the ignorance and anti-scientific superstition of the Dark Ages, the Islamic Middle East became a thriving center of science, human rights, and technological progress. Draper's unabashedly pro-Islamic portrayal of the history of science was eagerly adopted, translated, and promoted by the ideological founders of the Ottoman Turkish State and many in the West even referred to Draper as "the new apostle of Islamism."[57]

As Draper's 1874 book went through 50 printings and became an instant best-seller that was soon translated into multiple languages, the influence of his thesis was felt far and wide.[58] In time Draper and White's version of the history of science and religion reached axiomatic status at the popular level and

---

Italian, Dutch, Spanish, Polish, Japanese, Russian, Portuguese, and Serbian translations. Moore, *The Post-Darwinian Controversies*, 28.

55 John William Draper, *History of the Conflict between Religion and Science* (New York: Appleton, 1875), xi.

56 For example, Draper says that "Muslim astronomers' works had popularized the idea of a spherical Earth in the West and Columbus himself had been influenced by Averroes, but this view was deplored by Christian theologians as the Scriptures suggested a flat Earth," Draper, *History of the Warfare*, 160. Draper comments that because Christian doctrine holds that "the date of Creation was … not more than four or five thousand years before Christ" Christian theologians were "constrained to look with disfavor on the Mohammedan theory of the evolution of man from lower forms," Draper, *History of the Warfare*, 187–188.

57 "Draper's conflict between religion and science," *Catholic World* 21:122 (May 1875) 200–213, 178. For a discussion of Draper's influence on early Ottoman Turkish ideology see M. Alper Yalçinkaya, "Science as an ally of religion: a Muslim appropriation of 'the conflict thesis'" *British Journal for the History of Science* 44: 2 (2011); A cartoon was published in the *New York Ledger* (the widest-circulated publication of its day), showing Draper dressed as an Arab with turban, pointed shoes, and scimitar (engraved … By Science bright and *Islam's* might, tis I who tell thee so!" See Donald Fleming, *John William Draper and the Religion of Science* (Philadelphia: University of Pennsylvania Press, 1950), 135.

58 For example, in his diary Dostoevsky laments the impact of Draper's view "that Christianity was the source of ignorance and that science had been brought to the Christian world by the Mohammedans." Fyodor Dostoevsky, *A Writer's Diary*, vol. 2: 1877–1881, tr. Kenneth Lantz, (Evanston: Northwestern University Press, 1994), 1023. Draper also influenced Henry Steel Olcott and the reception of Buddhism as a "scientific religion"

"the Draper-White thesis, as it has come to be known, was enormously influential among academics."[59] During much of the twentieth century, the conflict thesis "dominated the historical interpretation of the relationship of science and religion,"[60] and by the 1930s "prevailing scholarly opinion, which had been shaped by the books of Draper and White, held that science and religion were inherently opposed and necessarily in conflict."[61] The continuing influence of Draper and White's narrative is attested by the fact that all modern readers are quite familiar with the basic outlines of their version of history and most would never even seriously consider contesting it.

Evaluating the Conflict Thesis

While the Draper-White warfare thesis remains immensely influential today, global popularity and continuing cultural influence are no substitutes for sound scholarship. And sound scholarship it is not. Draper frequently conjures up dramatic "historical" cases of conflict between science and religion while supplying no supporting citations from primary sources, and where he does cite sources, he often makes his case by taking quotes out of their original context (even to the point of making them say the opposite of the author's intention). Evaluating the scholarly value of Draper's work, historian of science Colin Russell points out that "Draper takes such liberty with history, perpetuating legends as fact, that he is rightly avoided today in serious historical study."[62] Historian of science Lawrence Principe agrees, explaining that Draper's book "is not good history: historical 'facts' are confected, and causes and chronologies are twisted to the author's purpose." Draper's thesis, says Principe, "rests on very shaky (and sometimes fabricated) foundations and was contrived largely for quite specific political, professional, and racist purposes."[63]

While White's work a bit less dramatic and has more of a historical feel, White is similarly guilty of regularly misconstruing his primary sources and

---

in the West. See Stephen Prothero, *The White Buddhist: The Asian Odyssey of Henry Steel Olcott* (Bloomington: Indiana University Press, 1996).

59 Gary Ferngren, "Preface," *The History of Science and Religion in the Western Tradition*, Edited by Gary B. Ferngren (New York: Garland, 2000), xii.

60 Ferngren, "Preface," *The History of Science and Religion in the Western Tradition*, xii.

61 David B. Wilson, "The Historiography of Science and Religion," in Gary Ferngren ed., *Science and Religion: A Historical Introduction* (Baltimore: Johns-Hopkins University Press, 2002), 4.

62 Colin Russell, "The Conflict of Science and Religion," in Gary Ferngren ed., *Science and Religion: A Historical Introduction* (Baltimore: Johns-Hopkins University Press, 2002), 15.

63 Lawrence Principe, "The Warfare Thesis," Science and Religion, recorded lecture (Chantilly, VA: The Teaching Company, 2006).

at times even sheer fabrication. "Though his prominent apparatus of prolific footnotes may create a misleading impression of meticulous scholarship," explains Russell, White's book "is no longer regarded as even a reliable secondary source for historical study."[64] Historian of science James Moore concurs, saying that "in the light of the confusion and misunderstanding perpetrated by the work … there is no reason to regard it as a scholarly book today, much less as an adequate interpretation."[65]

As a historical thesis, then, the notion of conflict between science and religion fails miserably. As Russell concludes, "serious historical scholarship has revealed the conflict thesis as, at best, an oversimplification and, at worst, a deception."[66] When the Draper-White thesis is examined closely, says historian of science Peter Harrison, "the historical record simply does not bear out this model of enduring warfare."[67] Consequently, concludes Principe, "serious modern historians of science have unanimously dismissed the warfare model as an adequate historical description."[68]

Even though their books of are of little or no value as historical scholarship, Draper and White's works can hardly be ignored due to their lasting impact on both the academy and popular culture. The Conflict or Warfare thesis has left its enduring mark. Capturing the imaginations of Draper and White's contemporaries, the idea of warfare between "science and Christian faith … has captivated writers on the subject, without respect of persons, for one hundred years."[69] The pernicious influence of the Draper and White thesis over both popular culture and the academy has been such that "one can hardly conduct business as usual when one is engaged in studying some historical episode relevant to both science and religion.'"[70] To accurately understand the relationship between science and religion, Draper and White cannot be ignored, and the extent of their thesis must be understood and their conclusions overcome. This is because, says, Numbers, "regardless of the scholarly merits of their works, they succeeded … in setting 'the terms of the debate' for future generations of historians."[71] Even among religious historians, says Moore, some of

64 Russell, "The Conflict of Science and Religion," 16.

65 Moore, *Post-Darwinian Controversies*, 40.

66 Russell, "The Conflict of Science and Religion," 16.

67 Peter Harrison, "Introduction," *The Cambridge Companion to Science and Religion*, 4.

68 Principe, "The Warfare Thesis."

69 Moore, *Post-Darwinian Controversies*, 40.

70 Maurice A. Finocchiaro, "Science, Religion, and the Historiography of the Galileo Affair: On the Undesirability of Oversimplification," *Osiris, Vol. 16, Science in Theistic Contexts: Cognitive Dimensions* (2001), 114–132, 115.

71 Numbers, "Science and Religion," 59.

them being apologists, "the majority, by falling prey to a captivating metaphor, have allowed Draper and White to 'set the terms of the debate.'"[72]

One can get a sense of the pervasive influence of Draper and White by considering the continuing popularity of the many myths that they constructed. Harrison explains, "Draper and White cataloged or constructed most of the now prevalent myths about the putatively pernicious influence of religion on science—medieval Christianity was inhospitable to science, the church banned human dissection, Pope Callixtus III excommunicated Halley's comet, ancient and medieval thinkers believed in a flat earth, heliocentrism was resisted because it demoted humans from the center of the cosmos, Galileo was tortured and imprisoned by the Inquisition, the church opposed vaccination and anesthesia, Darwin's *Origin* was met with universal resistance from religious quarters, and so on."[73] These persistent cultural memes are myths in two keys senses of the word—they are *foundational* to the validation of a specific worldview and a certain set of social practices, and they are historically *false*.[74] "Contrary to popular belief," explains historian of science James Hannam, "the Church never supported the idea that the earth is flat, never banned human dissection, never banned zero and certainly never burnt anyone at the stake for scientific ideas."[75] Galileo was never tortured and imprisoned by the Inquisition, and many of Darwin's most ardent advocates were deeply devout.[76]

72 Moore, *Post-Darwinian Controversies*, 43.

73 Harrison, *Territories*, 172. Harrison notes that these myths persist "in spite of the best efforts of historians of science." For a detailed criticism of the conflict thesis and a detailed account of why these myths are historically false see Lindberg and Numbers, *God and Nature*; and see also David Lindberg and Ronald Numbers, eds., *When Science and Christianity Meet* (Chicago: University of Chicago Press, 2003); Brooke, *Science and Religion*. Moore discusses the emergence and the impact of the conflict thesis in *The Post-Darwinian Controversies*.

74 See Joshua M. Moritz, *Science and Religion: Beyond Warfare and Toward Understanding* (Winona, MN: Anselm Academic, 2016).

75 James Hannam, *God's Philosophers: How the Medieval World Laid the Foundations of Modern Science* (Washington, D.C.: Regnery Publishing, 2011), xv.

76 Maurice A. Finocchiaro, "Myth 8: That Galileo Was Imprisoned and Tortured for Advocating Copernicanism," in *Galileo Goes to Jail, and Other Myths about Science and Religion*, Ronald L. Numbers, ed. (Cambridge, MA: Harvard University Press, 2009); David Livingstone, *Darwin's Forgotten Defenders: The Encounter Between Evangelical Theology and Evolutionary Thought* (Grand Rapids: Eerdmans, 1987).

From Conflict to Creative Tension and Creative Mutual Interaction

To understand why the Draper-White Conflict thesis is false requires some consideration of the actual historical interactions of science and religion. Yet, even speaking of the interactions *between* science and religion is misleading because such language is historically anachronistic. As Harrison explains, "much contemporary discussion about science and religion assumes that there are discrete human activities, 'science' and 'religion,' which have had some unitary and enduring essence that persists over time." However, he says "that this is not the case."[77] Throughout most of the history of science, the pursuit of scientific knowledge is motivated by religious concerns, and empirical investigations are done within a religious context. In the early modern period the pursuit of science is viewed as a form of worshipping God and the practitioners of science are often quite devout. As historian of science Kathleen Crowther explains, "Almost all early modern natural philosophers saw *what they* were doing as a pious activity because it involved exploring and contemplating the created works of God."[78] For these early scientists, "theological and scientific inquiry are not separate: to study the physical world means to study God's creation, to study God means to learn about the world."[79] The prominent scientists of the early modern period (who referred to themselves as natural philosophers) "did not distinguish what we would call the scientific aspects of their work from what we would call theology. Their study of the natural world was conceived as a study of God's creation, disclosing something of the nature of God."[80] For example, Robert Boyle, the founding father of modern chemistry, Niels Stensen, the founding father of stratigraphy, Johannes Kepler, Isaac Newton, and many other early scientists understand their scientific quest within the theological context of the Christian doctrine of creation. For Boyle, knowledge of nature illuminates the character of the Creator of nature. Rather than viewing religion and science as dichotomous entities, "Boyle had a remarkably unified worldview" and saw the truths of chemistry, physics and theology as "different aspects of one unified body of knowledge."[81] Kepler likewise sees his sci-

77 Harrison, *Territories*, 6.

78 Kathleen Crowther, "The Scientific Revolution" in *The Oxford Handbook of Early Modern European History, 1350–1750: Volume II: Cultures and Power*, ed. Hamish Scott (Oxford: Oxford University Press, 2015), 58.

79 Lawrence Principe, *The Scientific Revolution: A Very Short Introduction* (Oxford: Oxford University Press, 2011), 57.

80 John H. Brooke, "Preface," in *Science in Theistic Contexts: Cognitive Dimensions, Osiris 16*, ed. John H. Brooke, M.J. Osler, J.M. Van der Meer (University of Chicago Press, 2001), ix.

81 Jan Wojcik, *Robert Boyle and the Limits of Reason* (Cambridge: Cambridge University Press, 1997), ix; See also Michael Hunter, ed. *Robert Boyle Reconsidered* (Cambridge: Cambridge University Press, 1994).

ence as a theological quest. Indeed, says Principe, "Kepler became convinced of Copernicanism partly because the heliocentric universe provided a physical analogy to the Holy Trinity."[82] Because theology and science were so inextricably integrated during the period known as the Scientific Revolution there are many "well-known examples where theistic presuppositions appear to have done constructive work in the shaping of scientific theories."[83]

Within the Roman Catholic Church many early scientific investigations were carried out in monastic centers. The rural monasteries of Franciscans, Dominicans and Benedictines tended to become centers of natural investigation and technological innovation. The Jesuits, in particular, were engaged in establishing the foundations of science as they spearheaded the sciences of optics, magnetism, electricity, seismology and many other fields.[84] In fact, during the height of the scientific revolution the *religious* order of the Jesuits was the leading *scientific* organization in Europe.[85] As Jesuits spread across the globe, they developed the first global scientific correspondence network spanning India, China, and the Americas. The Jesuits "channelled everything from biological specimens and astronomical observations to cultural artefacts and extensive reports of native knowledge and customs back to Rome,"[86] and to other scientists in Europe. The Jesuits were great innovators in education during this time and they founded hundreds of new universities. "Jesuit schools were often the first to teach some of the new scientific ideas of the Scientific Revolution, and educated many of the thinkers responsible for them." "The Jesuit attitude in studies of science and mathematics expresses their motto

82 Principe, *Scientific Revolution*, 57.

83 Brooke, "Preface," ix.

84 "Many Jesuits worked to develop not only astronomy, but also optics and mechanics in physico-mathematical directions." Rivka Feldhay, "Religion," in *The Cambridge History of Science: Volume 3, Early Modern Science*, ed. Katharine Park and Lorraine Daston (Cambridge: Cambridge University Press, 2003), 739; John L. Heilbron, *Electricity in the 17th and 18th Centuries: A Study of Early Modern Physics* (Berkeley: University of California Press, 1979). See also *The Jesuits: Cultures, Sciences, and the Arts*, John W. O'Malley, Gauvin Bailey, Steven Harris, and Frank Kennedy, eds. (Toronto: University of Toronto Press, 1999).

85 Joshua M. Moritz, "Scientific Revolution," in *The Routledge Companion to Modern Christian Thought*, ed. Chad Meister and James Beilby (London; New York: Routledge, 2013), 268. In the 18th century, more than half the chairs of mathematics in Europe were held by Jesuits.

86 Principe, *Scientific Revolution*, 19. And scientific knowledge moved in the outbound direction as well. For instance, Johann Schreck was "a Jesuit missionary who brought European scientific knowledge to China." Principe, *Scientific Revolution*, 123. See also Michela Cigola, "The Influence of the Society of Jesus on the Spread of European Mechanical Knowledge in China in the XVIth and XVIIth Centuries," in *Explorations in the History of Machines and Mechanisms* (Amsterdam: Springer 2012), 69–79.

'to find God in all things'. While Jesuits emphasized this incentive, it was not unique to them—it undergirded virtually the entire Scientific Revolution."[87]

Human Dissection in the Middle Ages: A Case Study in Conflict or a Persistent Myth?

Recent accounts by professional historians of science have explored the numerous historical cases of harmony, mutuality, and symbiosis between science and religion. Since stories of conflict between science and religion loom large in the public imagination, however, here we will take a closer look at two frequently cited and alleged "cases of conflict"—the Church's attitude towards human dissection and the events surrounding the Galileo Affair—in order to demonstrate the failings of the warfare thesis and highlight the complex character of the interactions between science and religion.

A recent high-profile discussion of human dissection in the scientific journal *Anatomy & Cell Biology* states that, "Following widespread introduction of Christianity in Europe during the Middle Ages, the development of rational thought and investigation was paralysed by the church authorities ... During this period, human dissection was considered to be blasphemous and so was prohibited. For hundreds of years, the European world valued the sanctity of the church more than scientific quest ..."[88] Another recent article published in the *Croatian Medical Journal* states that "Since in the early Middle Ages, between [the] 9th and 11th century the development of rational thought and investigation was paralyzed by the Church authorities ... Anatomical dissection was illegal."[89] No citations are given for the primary sources of this information regarding the Medieval Church's denunciation of dissection as blasphemous, but if the authors were to trace the intellectual history of this idea they would find that it originates in the work of Andrew Dickson White. "The myth that the medieval church prohibited human dissection," explains historian of science Katherine Park, "was largely the creation of Andrew Dickson White in the late nineteenth century."[90] White argues that the Church was opposed

87 Principe, *Scientific Revolution*, 19.

88 Sanjib Kumar Ghosh, "Human cadaveric dissection: a historical account from ancient Greece to the modern era" *Anatomy & Cell Biology* 48:3 (September 2015). The visibility of this article is reflected in the fact that it is cited in the Wikipedia entry on "Dissection". For the current Wikipedia article on Dissection see https://en.wikipedia.org/wiki/Dissection.

89 Alexandra Mavrodi and George Paraskevas, "Mondino de Luzzi: a luminous figure in the darkness of the Middle Ages" *Croatian Medical Journal* 55:1 (February 2014), 50–3.

90 Katherine Park, "Myth 5: That the medieval church prohibited human dissection," in *Galileo Goes to Jail: And Other Myths about Science and Religion*, Ronald L. Numbers, ed. (Cambridge: Harvard University Press, 2009), 43–49.

to human dissection from the very beginning and, unlike the authors of the two scientific journals referenced above, White does cite a number of primary sources to show how the early church and the Medieval church were rabidly opposed to both dissection and the study of anatomy. To show how the early church rejected the study of anatomy and the practice of dissection White says that "St. Augustine held anatomy in abhorrence" and as his evidence he quotes Augustine in the following passage:

> I pass, now, to fields of more immediate importance to us—to anatomy and medicine. It might be supposed that the votaries of sciences like these would be suffered to escape attack; unfortunately, they have had to stand in the thickest of the battle. The Church, even in its earliest centuries, seems to have developed a distrust of them ... St. Augustine, in his great treatise on the City of God, which remains to this day one of the treasures of the Church, speaks with some bitterness of 'medical men who are called anatomists,' and says that 'with a cruel zeal for science they have dissected the bodies of the dead, and sometimes of sick persons, who have died under their knives, and have inhumanly pried into the secrets of the human body to learn the nature of disease and its exact seat, and how it might be cured!' But it was not until the mixture of theology and science had begun to ferment, in the thirteenth century, and the ecclesiastical power had been aroused in behalf of this sacred mixture, that the feeling against medical science broke into open war.[91]

It is unfortunate, considering the impact of White's work on later generations, that White takes Augustine violently out of context. If one looks at the larger context of Augustine's discussion one will see that Augustine is actually making quite the opposite point and he even provides a theological motivation for the practice of dissection:

> Assuredly no part of the body has been created for the sake of utility which does not also contribute something to its beauty. And this would be all the more apparent, if we knew more precisely how all its parts are connected and adapted to one another, and were not limited in our observations to what appears on the surface; for as to what is covered up and hidden from our view, the intricate web of veins and nerves, the vital parts of all that lies under the skin, no one can discover it. For although, *with a cruel zeal for science, some medical men, who are called anatomists,*

91 White, *A History of the Warfare*, 99–100.

have dissected the bodies of the dead, and sometimes even of sick persons who died under their knives, and have inhumanly pried into the secrets of the human body to learn the nature of the disease and its exact seat, and how it might be cured, yet those relations of which I speak, and which form the concord or, as the Greeks call it, 'harmony,' of the whole body outside and in, as of some instrument, no one has been able to discover, because no one has been audacious enough to seek for them. But if these could be known, then even the inward parts, which seem to have no beauty, would so delight us with their exquisite fitness, as to afford a profounder satisfaction to the mind—and the eyes are but its ministers—than the obvious beauty which gratifies the eye.

AUGUSTINE OF HIPPO, *The City of God, XXII*

But what about the Church after Augustine? White continues his argument lamenting how the Medieval Church "forbade medicine" and "interdicted dissection as sacrilege." He cites the 1248 Council of Le Mans "*The church abhors the shedding of blood*" as a unilateral prohibition of human dissection and comments that "Many other councils did the same."[92] White singles out Pope Innocent III as a particularly resilient bastion against both the practice of dissection and the administration of medicine and he states that "the decretal of Pope Boniface VIII was universally construed as forbidding all dissection." The Church, according to White denounced those who practiced medicine as "sorcerers", "leaguers with the devil" and "Mohammedans."[93] White tells his readers that Vesalius, "braving the fires of the Inquisition," studied the science

92 White, *A History of the Warfare*, 31; In the journal *Popular Science* Andrew Dickson White adds, "this maxim was used with deadly effect against the progress of surgery," *Popular Science Monthly*, Part I: 39 (May 1891); As an example of how widespread this prohibition is cited see the entry on "Healing," in the *New Dictionary of Christian Ethics & Pastoral Theology*, David Atkinson and David Field, eds. (Downers Grove, IN: InterVarsity Press, 1995), 486.

93 White, *A History of the Warfare*, 101–102; White's reference to Muslims as practitioners of human dissection is mistaken. There is no evidence that medieval Muslims practiced human dissection. "If indeed any medieval Islamic physician did undertake dissection, he must have felt the need to remain quiet about the details.... The silence of the medical community regarding the details of dissection, combined with the lack of concern regarding dissection in the legal literature (neither condemning nor approving nor even mentioning it), implies a lack of activity." On the other hand, "unambiguous and uncritical rendering of references to human dissection implies that, in ninth-century Baghdad, at least among the community of Christian physicians, the idea may not have been repugnant or unthinkable." Emilie Savage Smith, "Attitudes Toward Dissection in Medieval

of anatomy "despite ecclesiastical censure, great opposition in his own profession, and popular fury" and "in the search for real knowledge he risked the most terrible dangers, and especially the charge of sacrilege, founded upon the teachings of the Church for ages."[94]

Is White's claim about the medieval church historically accurate? Not in the least. There is no historical evidence that the church ever banned or discouraged the study of anatomy or the practice of human dissection. The purported prohibition of the Council of Le Mans that White cites was a literary fiction fabricated by the 18th century French surgeon and amateur historian François Quesnay. As Park points out, the "prohibition, *The church abhors the shedding of blood*, which White described as promulgated at the Council of Le Mans in 1248, was shown forty years ago to be a 'literary ghost,' produced by an inept eighteenth century French historian."[95]

The 1299 bull of Boniface VIII, *Detestande feritatis*, that White refers to was not directed against the study of anatomy, human dissection, or autopsy but rather was aimed at preventing a funerary practice that had become common among aristocrats and nobles of northern Europe who died while crusading in foreign lands. As Park explains, "Boniface's decree condemned not dissection or autopsy but a much more extreme funerary practice that involved dismembering the body and boiling the flesh off the bones in order to allow them to be more easily transported for distant burial."[96] Boniface's proclamation did not have an adverse effect on the study of anatomy or discourage the practice of dissection. "From all available evidence, Boniface's bull and letter were taken as irrelevant by generations of Italian medical professors, private doctors, judges, city councils, and even by later popes, several of whom were themselves embalmed."[97]

Was Innocent III an adamant opponent of the study of anatomy or the practice of dissection? There is no evidence that would indicate this was the case. In fact, explains historian of science James Hannam, "Pope Innocent III is on record ordering the forensic examination of a murder victim. These

---

Islam," *Journal of the History of Medicine and Allied Sciences* 50: 1 (1 January 1995), 89 and 105.

94 White, *A History of the Warfare*, 50.

95 Park, "Myth 5: That the medieval church prohibited human dissection," 46; see also Louis Bakay, *Neurosurgeons of the Past* (Illinois: Charles Thomas, 1987), 47; Renate Blumenfeld-Kosinski, *Not of Woman Born: Representations of Caesarean Birth in Medieval and Renaissance Culture* (Ithaca, N.Y.: Cornell University Press, 1990), 25.

96 Katharine Park, "The Criminal and the Saintly Body: Autopsy and Dissection in Renaissance Italy," *Renaissance Quarterly* 47.1 (1994) 1–33, 10.

97 Park, "Autopsy and Dissection in Renaissance Italy," 11.

autopsies broke the taboo against cutting up human bodies and, shortly afterwards, the medical faculty in Bologna was carrying out dissections as part of doctors' training."[98] Thus, says historian of medicine Ynez Viole O'Neill, "the first recorded case of a medico-judicial dissection … which was conducted at Bologna in February 1302, may be viewed therefore as representing not a break with tradition, but the outgrowth of a burgeoning development whose roots were firmly imbedded in both Canon and Civil law. By encouraging a shift of vision, Innocent III … sponsored a process with far-reaching consequences for the evolution of anatomy and of the other biological sciences."[99] Far from inhibiting the development of the science of anatomy, Innocent III exhibited "a spirit of inquiry … which helped to supply the impetus necessary to inaugurate the acceptance of scientific post-mortem examinations from which academic dissections, and ultimately the modern study of human anatomy evolved."[100]

What about Vesalius having to live in constant fear of the inquisition and his being persecuted by the church for practicing science? Park says that there is certainly no evidence "that Vesalius ran afoul of church authorities."[101] Hannam likewise affirms that Vesalius "never got into trouble for practicing human dissection" and that he "suffered from no religious persecution." In fact, says Hannam, Vesalius was "a man of great piety who saw his life's work as a way of glorifying God." When Vesalius published his great treatise on human dissection "the entire work was a paean to the magnificent handiwork of the Creator as uncovered by his servant."[102]

In contrast to White and the contemporary promulgators of the myth that the Church banned dissection, Park explains that "most medieval church authorities not only tolerated but encouraged the opening and dismemberment of human corpses to religious ends."[103] Park says there are no known cases "in which an anatomist was ever prosecuted for dissecting a human cadaver and no case in which the church ever rejected a request for a dispensation to dissect."[104]

If it wasn't the adversarial influence of Christianity that is to blame, why then was there such a long hiatus between the first recorded practices of

98 Hannam, *God's Philosophers*, 253.

99 Ynez Violé O'Neill, "Innocent III and the Evolution of Anatomy," *Medical History*, 20 (1976): 429–433, 432.

100 O'Neill, "Innocent III and the Evolution of Anatomy," 432.

101 Park, "Myth 5: That the medieval church prohibited human dissection," 47.

102 Hannam, *God's Philosophers*, 256.

103 Park, "Myth 5: That the medieval church prohibited human dissection," 44.

104 Park, "Myth 5: That the medieval church prohibited human dissection," 47.

human dissection in ancient Alexandria in 300 BCE and those that took place in Italy 1500 years later?[105] Park explains that "avoidance of dissection seems to have reflected preemptive caution on the part of anatomists rather than actual ecclesiastical pressure." It was not the activity or theology of the church that inhibited the study of anatomy but rather a combination of practical and social factors that have existed in numerous human cultural contexts throughout time. First and foremost among factors that discouraged any culturally respectable practice of dissection were the culturally ubiquitous and religiously pervasive taboos against touching corpses. In Roman religion, African Igbo religion, Judaism, Hinduism, and many other religious traditions the touching of a corpse results in ritual impurity or uncleanliness. Living in the Roman Empire in the 2nd century CE the physician Galen's limited access to human anatomy was "the result of a taboo against touching corpses and cutting into bodies."[106] As the Roman Empire transitioned into Christian Europe the limited availability of cadavers for dissection and the study of anatomy "resulted not from religious prohibitions but from strong cultural sensitivities concerning who was and was not an appropriate subject for dissection, sensitivities that were enshrined in local law." Even when local laws permitted human dissection, insecurities of a more personal nature kept cadavers in short supply. As Park explains, "To be exhibited naked in front of a group of university students—augmented, in the sixteenth century, by local notables and visiting dignitaries—was a deeply shaming prospect, particularly since dissection rendered the body unsuited for a normal funeral."[107] Even in the 16th century, well after the practice of dissection had become a standard part of medical education "a chronic shortage of cadavers … led enterprising students to violate graves and even … to hijack corpses from funeral processions." Given the fact that most societies and cultures did have a strong aversion to the practice of human dissection "it is surprising," says Hannam, "that the Church allowed dissections to go ahead with barely a whimper of opposition." Beyond just

105 "The first documented formal dissection of this sort was recorded in Padua in 1341, though it is clear that the practice was considerably older." Paula Findlen, "Anatomy theaters, botanical gardens, and natural history collections," in *The Cambridge History of Science: Volume 3, Early Modern Science*, ed. Katharine Park and Lorraine Daston (Cambridge University Press, 2003), 275.

106 Margaret J. Osler, *Reconfiguring the World: Nature, God, and Human Understanding from the Middle Ages to Early Modern Europe* (Baltimore: Johns Hopkins University Press, 2010), 27.

107 "Families had no qualms about autopsies, which were becoming increasingly common in this period, since they were performed privately and left the body substantially intact for the funeral procession," Park, "Myth 5: That the medieval church prohibited human dissection," 48.

toleration, others have shown that in a number of key ways the Church "supplied the impetus necessary to initiate the study of human anatomy nurtured thereafter in the university, a distinctive product of Western Europe." This first case study in the alleged conflict between science and religion, then, dissipates in light of further examination. Instead of conflict one finds a complex interaction of cultural, social, and theological factors where the church can be seen as having played a broadly supportive role in the emergence of a particular scientific practice.

The Galileo Affair: The Smoking Gun in the War between Science and Religion?

A second oft-cited case study in conflict between science and religion is the Galileo affair. The intellectual issues, politics, conflicting personalities, and circumstances surrounding the trial of Galileo are complex. As Principe explains, "the events are tied up with offended feelings, political intrigue, who was qualified to interpret Scripture, being at the wrong place at the wrong time, and being caught between Church factions."[108] Some grasp of the historical complexity is necessary, however, if one is to understand what actually happened and what was at stake. One thing that is clear is that the Galileo affair "was not a simple matter of 'science versus religion.'"[109] To cite the Galileo affair as "an instance of science-religion conflict," says Harrison, "is to misconstrue the context."[110] Both before and after his trial—Galileo was a practicing and believing Christian who affirmed "that there could be no fundamental opposition between God's words and His works" and "throughout his life he functioned as the good Catholic he claimed to be."[111] The Church funded Galileo's research and the large salary Galileo enjoyed as court mathematician and philosopher came from ecclesiastical revenues. Throughout his scientific career he "enjoyed the respect, trust, and protection of many high Church officials,"[112] and he received a papal pension, becoming "a cleric in the summer of 1630 to the extent of wearing a tonsure and reciting a daily office."[113] Some historians have even argued "that Galileo was a zealot *not* for the Copernican astronomy,

108 Principe, *Scientific Revolution*, 61.

109 Principe, *Scientific Revolution*, 61.

110 Harrison, *Territories*, 172.

111 John Heilbron, *Galileo* (Oxford: Oxford University Press, 2010), 364.

112 Maurice A. Finocchiaro, "The Copernican Revolution and the Galileo Affair," in *The Blackwell Companion to Science and Christianity*, ed. J.B. Stump and A.G. Padgett (Chichester: Wiley, 2012), 18.

113 Heilbron, *Galileo*, 299.

but for the future of the Catholic Church and for the protection of religious faith against *any* scientific discovery that might be made."[114] During his trial Galileo received "unprecedentedly benign treatment" and was permitted to lodge at the palatial Tuscan embassy and for a few days at prosecutor's six-room apartment, together with a servant, who brought him meals twice a day from the Tuscan embassy.[115] Galileo was never tortured or imprisoned. And long after his trial when he was old and blind and confined to his villa, members of religious orders comforted and read to him.

Understanding the factors that led up to Galileo's trial requires some understanding of the context of Galileo's scientific work and the intellectual history immediately preceding it. The idea that the Earth might be moving was a subject of theoretical discussion long before Galileo published his argument for Earth's motion from the tides in 1632. John Buridan (c.1300–c.1361), rector of the University of Paris, seriously considered the possibility of Earth's motion and argued that the apparent turning of the night sky might just as well be an illusion caused by the Earth's rotation.[116] Buridan's student, Bishop Nicole Oresme (c.1325–82), went further and challenged Aristotle's theory of motion by explaining that if the Earth is indeed rotating, than an arrow fired into the air has two types of motion—earth's rotational motion plus the motion imparted to the arrow by the bow. Oresme concluded, however, that we cannot use either reason or observation to conclusively determine if the earth moves. "All motions are relative," explains Oresme, "and without a point of reference it is impossible to decide whether the Earth is spinning or the heavens revolving."[117] Indeed, he says, "only an extra-terrestrial view of the earth could determine the truth about its motion." Cardinal Nicholas of Cusa (1400–64) likewise follows Buridan in arguing for the Earth's motion and says that we do not notice its movement because we are riding along with it. Beyond this Cusa "argued that in order to reflect God's majesty, the universe he created would have to be limitless, if not quite infinite ... Therefore," says Cusa, "the earth cannot be in the center ... and just as the earth is not at the center of the universe, so the sphere of the fixed stars is not its outer border."[118] Cusa even hypothesizes that the Earth is just another star and suggests that alien life forms could exist elsewhere in the universe. Cusa, an orthodox Roman Catholic throughout

114 Stillman Drake, *Galileo: A Very Short Introduction* (New York: Oxford University Press, 2001), xix.

115 Finocchiaro, "Myth 8: That Galileo Was Imprisoned and Tortured," 74.

116 Hannam, *God's Philosophers*, 179, 275.

117 Principe, *Scientific Revolution*, 50.

118 Quoted in Hannam, *God's Philosophers*, 194.

his whole life, and even the head of the church for a time, was never the subject of controversy and his theories about the natural world were in no way viewed as heretical. This is because both before Cusa and afterwards there was much freedom to theorize and speculate about such matters, and—as long as such speculation stayed hypothetical—no one would have objected or raised any concerns.

When Nicolas Copernicus (1473–1543) first proposed his heliocentric theory of the Earth's motion in 1514, he was doing so within the context of a significant amount of intellectual freedom, and his ideas about Earth's motion belonged to a long and distinguished pedigree. The case of Copernicus was certainly not an essential conflict between science and religion and the church even played a proactive role in bringing his theory to light. Copernicus did his scientific work while serving as a canon in the Cathedral of Varmia and his fellow canons and numerous other church officials actively encouraged him to pursue and publish his findings. For example in November of 1536, Cardinal Nicolas von Schoenberg wrote to Copernicus saying: "I have learned that you teach that the Earth moves; that the Sun occupies the lowest, and thus the central place in the world ... and that you have prepared expositions of this whole system of astronomy ... Therefore with the utmost earnestness I entreat you, most learned sir, unless I inconvenience you, to communicate this discovery of yours to scholars."[119] Why then, was Copernicus so reticent to publish a complete exposition of his system? The answer is that he was being cautious due to concerns about the criticism he would receive from *fellow astronomers*. Anxious about how his theory would be received by his scientific peers Copernicus "kept refining it for over 25 years, and had it not been for the nagging of several prominent churchmen it might never have been published."[120]

Copernicus was finally convinced to publish a full manuscript of his theory after the Wittenberg astronomer Georg Joachim Rheticus published a summary of Copernicus's ideas that was well received. When Copernicus' *On the Revolutions of the Heavenly Spheres* was printed "the book's appearance did not unleash the criticism Copernicus feared."[121] As historian of science Owen Gingerich explains, "the immediate response to Copernicus's *Revolutions* was mild. No serious attacks were mounted against it; nor did it receive any immediate flurry of support. Scholars and libraries across Europe purchased the five

119 Quoted in Owen Gingerich and James MacLachlan, *Nicolaus Copernicus: Making the Earth a Planet* (Oxford: Oxford University Press, 2005), 98; and in Principe, *Scientific Revolution*, 49.

120 Principe, *Scientific Revolution*, 49.

121 Principe, *Scientific Revolution*, 50.

hundred printed copies of *Revolutions*. It seems that every astronomer who took his own work seriously wanted a copy. A little more than 20 years later, the book was reprinted in Basel, Switzerland, and spread more widely. During the latter half of the 1500s, the name Copernicus often appeared in astronomical texts and lectures."[122] Although his book was widely read and discussed, Copernicus's heliocentric system did not fit the observational data any better than the geocentric system and thus few of his scientific peers were convinced. For empirically minded thinkers the notion of a moving Earth conflicted with both basic physics and common sense. If the theory of Copernicus were true "then the human senses would be lying to us," and many "regarded as absurd that the senses should deceive us about such a basic phenomenon as the state of the terrestrial globe on which we live."[123] In other words, "the geokinetic theory seemed to be in flat contradiction with direct sense experience, and so to violate the fundamental epistemological principle claiming that under normal conditions the senses provide us with an access to reality."[124] Moreover, the astronomers of the time knew that "if the Earth moves around the Sun, the stars should exhibit parallax—a small shift in their apparent relative positions as the Earth swings from one side of its orbit to the other. But no parallax could be detected." This meant "that either the Earth was *not* moving, or the stars were *incomprehensibly* far away ... Copernicus estimated Saturn's sphere to be about 40 million miles away, but the lack of stellar parallax meant ... that the stars would have to be at least 150 *billion* miles farther out."[125] While Copernicus had a theologically elegant, philosophically harmonious, and mathematically robust theory, he had not developed the new physics that was needed to support his theory and he had no *empirical proof* by which he could demonstrate the Earth's motion.

Thirty years after Copernicus published his *Revolutions* the Danish Lutheran astronomer Tycho Brahe made a series of observations that eventually led him to reject the Ptolemaic model of the solar system that was based on the geocentric cosmology of Aristotle. Observing comets in 1577 and 1585 which, contra Aristotle's theory, appeared to be located far above the sphere of the moon and orbiting the sun, Tycho abandoned the Aristotelian-Ptolemaic idea that each of the planets was carried by a solid crystalline orb. Instead he "concluded

122 Gingerich and MacLachlan, 108–109.

123 Maurice A. Finocchiaro, *Defending Copernicus and Galileo: Critical Reasoning in the Two Affairs* (Springer, 2010), 25.

124 Maurice A. Finocchiaro, *The Routledge Guidebook to Galileo's Dialogue* (Routledge, 2013), 35.

125 Principe, *Scientific Revolution*, 50.

that the substance of the heavens was a continuous fluid of some sort, that the planets moved freely through this medium, and that the orbs of the planets were not physical objects but geometrical constructions representing boundaries in this medium."[126] Convinced by the common sense observation that the Earth is not moving Tycho constructed an alternative model of the solar system which placed the Sun at the center of the orbits of Mercury, Venus, Mars, Jupiter and Saturn while the orbits of the Sun and the Moon revolved around a stationary Earth at the center of the cosmos. According to many of his contemporaries Tycho had "proven clearly ... the fact that for Saturn, Jupiter, Mars, Venus, and Mercury the center of the earth is not their center."[127] Tycho's introduction of a geo-heliocentric cosmic scheme in 1588 meant that the Copernican theory of the solar system was no longer a unique alternative to the geocentric Ptolemaic model. "There were now two possible explanations for all the phenomena claimed as evidence for Copernicus and against Ptolemy."[128]

Lutheran astronomer Johannes Kepler became Tycho's assistant in 1600 and then was appointed his successor as Danish Royal Astronomer when Tycho died a year later. In that same year Kepler remarked that "today there is practically no one who would doubt what is common to the Copernican and Tychonic hypotheses, namely that the sun is the center of motion of the five planets."[129] Kepler had become a convinced Copernican in the 1590s owing to both empirical and theological reasons and his scientific "method was grounded in the conviction that God had created the universe according to 'number, weight and measure' and therefore according to mathematical laws."[130] Having found the cause for the number and distances of planets, and uncovering "a geometrical structure to the heavens whose elegant beauty served as the best proof yet of the Copernican system," Kepler believed that he "had uncovered

126 Peter Barker and Bernard R. Goldstein, "Theological Foundations of Kepler's Astronomy," in *Science in Theistic Contexts: Cognitive Dimensions, Osiris 16*, ed. John H. Brooke, M.J. Osler, J.M. Van der Meer (University of Chicago Press, 2001), 93.

127 Noah J. Efron and Menachem Fisch, "Astronomical Exegesis An Early Modern Jewish Interpretation of the Heavens," in *Science in Theistic Contexts: Cognitive Dimensions, Osiris 16*, ed. John H. Brooke, M.J. Osler, J.M. Van der Meer (University of Chicago Press, 2001), 74.

128 Barker and Goldstein, 92.

129 Nicholas Jardine, *The Birth of History and the Philosophy of Science: Kepler's A Defense of Tycho against Ursus with Essays on its Provenance and Significance* (Cambridge: Cambridge University Press, 1984), 147.

130 Ann Blair, "Natural Philosophy," in *The Cambridge History of Science: Volume 3, Early Modern Science*, ed. Katharine Park and Lorraine Daston (Cambridge: Cambridge University Press, 2003), 401.

the mathematical blueprint by which God constructed the heavens."[131] In the year 1600, Tycho had directed Kepler to study the motions of Mars and, after a long struggle to assign it a circular path that fit the positions that Tycho observed, Kepler came to the startling conclusion that the planet's positions could best be accounted for best by an *elliptical* orbit rather than a circular orbit. For two thousand years, all astronomers—including Copernicus and Galileo—had held to the circular motion of the planets and as Kepler reluctantly broke with this age-old astronomical tradition he discovered his famous three laws of planetary motion.[132] To explain the elliptical nature of the planetary orbits Kepler drew on a recent claim of English physicist William Gilbert (1544–1603) that the Earth is a gigantic magnet. Kepler postulated that the Sun likewise exerted a force analogous to magnetism, which attracted the planets and decreased in proportion to distance. Kepler also suggested that the moon exerted the same type of attractive force upon the Earth and that this force was responsible for the tides.

By the time Galileo began his scientific career, "all the evidence was running one way; if not directly *for* Copernicus, then at least directly *against* Aristotle-Ptolemy."[133] Although the model of Copernicus was well known and even taught as an alternative to the Ptolemaic system, for many astronomers "the Tychonic system was more weighty in the scales of evidence that the Copernican."[134] By the second decade of the 1600s the astronomers at the leading Jesuit University, the Collegio Romano, had abandoned the Aristotelian geocentric model of Ptolemy for the geo-heliocentric Tychonic system, and

131 Principe, *The Scientific Revolution*, 57.

132 This move to elliptical orbits "made Kepler's work more revolutionary for technical astronomy than had been the cosmological innovation of Copernicus. If we look at the strictly scientific aspect of the Copernican revolution, Kepler's contribution was much greater than Galileo's. Working astronomers had no choice but to use his ellipses; this model incidentally resolved some classic problems of planetary astronomy (as that of planetary 'latitudes'); and through Kepler's work the Copernican system became the working hypothesis for astronomy." J.R. Ravetz, "The Copernican Revolution," in *The Companion to the History of Modern Science*, ed. G.N. Cantor, J.R.R. Christie, M.J.S. Hodge, R.C. Olby (London: Routledge, 1996), 210. William Shea and Mariano Artigas comment that "to the end of his life, Galileo held to a simplified version of the Copernican system in which all the planets move in perfect circles. Although he preached openmindedness, he never lent an ear to Kepler's arguments about elliptical paths." William R. Shea and Mariano Artigas, *Galileo in Rome: The Rise and Fall of a Troublesome Genius* (Oxford: Oxford University Press, 2003), 26.

133 Ravetz, "The Copernican Revolution," 213.

134 Ravetz, "The Copernican Revolution," 213.

Tycho's system was taught in Jesuit astronomy texts. The "dismissive attitude towards Ptolemaic astronomy—treating as beyond dispute that at least the five Tychonic planets orbit the Sun—had held for many years" before Galileo turned his telescope to the sky in 1609, and this general skepticism regarding the Ptolemaic system predates Galileo's 1610 discovery of the phases of Venus.[135] The question of the Earth's mobility was the only clear point that divided Tycho's model from that of Copernicus and this issue thus came to be a primary area of scientific contention in the time of Galileo.[136]

Jesuit Cardinal Robert Bellarmine, a key player in the Galileo affair, "had rejected Aristotle's natural philosophy" (in part, due to its lack of Biblical support) and he was equally "relaxed about the collapse of Ptolemy's astronomy."[137] Bellarmine's rejection of the Ptolemaic model, however, was not tantamount to his accepting the Copernican system as a physical fact. Bellarmine, and all the astronomers of his day knew that other models—such as that of Tycho's—were viable interpretations of the available data. The Catholic Church as a whole was scientifically cautious. As Harrison explains, the "Church endorsed the *scientific* consensus of the period, which, on the basis of the available evidence, held that the earth was stationary in the middle of the universe."[138] Moreover, *within* the church of Galileo's day there was much debate about these scientific matters, where the Dominicans represented the scientifically *conservative* or cautious side of the debate and the Jesuits represented the scientifically *progressive* or speculative side. Dominican scientists, who were the traditional intellectual elite of the Church "developed epistemological scepticism and tended to view Copernicanism not only as an *unproved* theory but also as an *unprovable* one," and held to a type of scientific instrumentalism where it concerned mathematical astronomy.[139] In contrast, the Jesuit scientists—the institutional rivals of the Dominicans who were fast becoming the new intellectual elite of the church—had begun, under the leadership of mathematician and astronomer Christoph Clavius (1537–1612), to advocate scientific realism in mathematical astronomy. Galileo agreed with the

135 Howard Margolis, "Tycho's system and Galileo's *Dialogue*," *Studies in History and Philosophy of Science* 22:2 (1991), 259–275, 264.

136 Principe, *Scientific Revolution*, 59.

137 Hannam, *God's Philosophers*, 322–323.

138 Harrison, *Territories*, 172.

139 Rivka Feldhay, *Galileo and the Church: Political Inquisition Or Critical Dialogue?* (New York: Cambridge University Press 1995), 10; Though they were scientific realists with regard to other areas of natural philosophy, such as physics, Ernan McMullin, "Galileo on Science and Scripture," in *The Cambridge Companion to Galileo*, ed. Peter Machamer (Cambridge: Cambridge University Press, 1998), 271–347, 329.

Jesuit realist approach to science and disagreed with the instrumentalism of the Dominicans.[140] Thus, in the early phase of Galileo's scientific career the Dominicans were his critics and the Jesuits were his defenders.[141]

When Galileo began his physical investigations in 1586 "he was acutely aware of the strength of the observational astronomical evidence against Copernicanism" and "until 1609 Galileo judged that the anti-Copernican arguments outweighed the pro-Copernican ones."[142] Galileo was "attracted to Copernicanism because he felt the earth's motion could best explain why the tides occur. But he had not published or articulated this general intuition and this particular feeling."[143] Given the context of the ongoing debate between Jesuit and Dominican scientists, philosophers, and theologians, Galileo's discovery of mountains on the moon, the moons of Jupiter, and the phases of Venus in 1610 did not lead either to his persecution or to any great theological upheaval in the Church. Cardinal Bellarmine asked the Jesuit astronomers at Rome whether they could confirm Galileo's observations. When they finally obtained a telescope powerful enough to permit confirmation, the Jesuits replied that they indeed saw what Galileo had described, and they went on to make several new discoveries of their own. While the Jesuit astronomers who confirmed Galileo's observations "were not sure about how the results should be interpreted," they supported Galileo and his understanding of the data.[144] When Galileo visited Rome in 1611, "he received a hero's welcome and the Jesuits lauded him openly for his wonderful discoveries."[145] The Jesuit astronomers in Rome gave a feast in Galileo's honor and Clavius, the senior member of the Collegio Romano and one of the most respected mathematicians in Europe wrote that "Galileo's discoveries required a rethinking of the structure of the heavens."[146] Galileo was elected to the exclusive scientific society the Accademia dei Lincei, support for him spread, and he was soon an acclaimed

140 William Wallace demonstrates that there was "a direct line of influence running from Jesuits who taught natural philosophy and mathematics to Galileo's experiments and publications on free-fall. Moreover, Wallace shows that the method that guided Galileo's experiments, the so-called demonstrative *regressus*, depended directly on the Roman Jesuits' Aristotelian commentaries." John W. O'Malley, "Introduction to Part III," in *The Jesuits II: Cultures, Sciences, and the Arts, 1540–1773*, ed. John W. O'Malley (Toronto: University of Toronto Press, 2006), 288.

141 Feldhay, *Galileo and the Church*, 102.

142 Finocchiaro, *Defending Copernicus and Galileo*, xvi.

143 Finocchiaro, "The Copernican Revolution and the Galileo Affair," 17.

144 Hannam, *God's Philosophers*, 319.

145 Hannam, *God's Philosophers*, 319.

146 Principe, *Scientific Revolution*, 61.

celebrity. Galileo's fame increased in 1612 when the Accademia published his book *Letters on Sunspots*. In the *Letters*, Copernicus's "heliocentric theory is treated as a fact."[147] Galileo framed his argument in the *Letters* so as to suggest that the phases of Venus defeated the only available geocentric model and consequently established the validity of Copernicus's heliocentric alternative "with absolute necessity."[148]

Galileo's rapid ascendency to prominence—especially when combined with his quick wit, which he often used to ridicule his opponents—made him the focus of much enmity and resentment. "Galileo's aim was not to write carefully reasoned scholarly papers of the sort astronomers and cosmologists now write, but to influence public opinion and win the cosmological debate in the public arena."[149] His rhetorical appeals to public opinion and his increasing insistence that Copernicus was the only viable option alienated his more cautious scientific supporters who were not convinced. For them, Tycho Brahe's model still made the most sense.[150]

In 1613–4 Galileo entered more deeply into the discussion over Copernicanism through a widely circulated letter to his disciple—a Benedictine monk and professor of physics at Pisa—Benedetto Castelli. In December 1615 Galileo went to Rome—of his own accord—to defend the Copernican system as a physical reality. While in Rome, Galileo, "was able to talk to many influential Church officials and was received in a friendly manner."[151] In the same year the Dominican friar Niccolò Lorini filed a written complaint against Galileo with the Inquisition in Rome, and forwarded a copy of Galileo's letter to Castelli. Another Dominican made a formal deposition in person against him. The Inquisition found no serious objections in Galileo's letter and dismissed both complaints against him. Galileo was never formally charged, nor even informed of these proceedings and the case was closed. Commenting on these events, historian of science, Stillman Drake explains, "It is apparent

---

147 Noel M. Swerdlow, "Galileo's Discoveries with the Telescope and their Evidence for the Copernican Theory," in *The Cambridge Companion to Galileo*, ed. Peter Machamer (Cambridge: Cambridge University Press, 1998), 244–270, 246.

148 David C. Lindberg, "Galileo, the Church, and the Cosmos," in *When Science and Christianity Meet*, David C. Lindberg and Ronald L. Numbers, eds. (Chicago: University of Chicago Press, 2003), 33–60, 43.

149 Lindberg, "Galileo, the Church, and the Cosmos," 44.

150 On top of this, a controversy between Galileo and the Jesuit astronomer Christopher Scheiner over the priority of discovering sunspots resulted in a cooling of relations between Galileo and his Jesuit friends and supporters. See Lindberg, "Galileo, the Church, and the Cosmos," 45.

151 Finocchiaro, *Routledge Guidebook to Galileo's Dialogue*, 43.

that theologians were not seeking some pretext to censure Galileo, let alone to intervene in scientific issues. The problem was rather with intriguing personal foes of Galileo, and an ambitious priest, than with responsible Church officials."[152]

Because the issue of Copernicanism kept arising in a number of contexts, the Inquisition commissioned a panel of consultants to examine the geokinetic thesis and report on its philosophical and theological merits. The investigation lasted about a year and in February 1616, the panel concluded that Copernicanism was "foolish and absurd" in philosophy and "heretical" in theology. However, the Inquisition took no action on the panel's recommendation and did not issue the formal condemnation of Copernicanism recommended by the consultants. One of the main reasons why the Inquisition did not endorse the recommendations of the advisory panel is because at the March 3, 1616 meeting two prominent Cardinals—Bonifazio Caetano and Maffeo Barberini—strongly objected. Campanella was not a Copernican, "but he wanted to preserve the intellectual freedom of Catholic scientists."[153] Barberini was well educated Florentine who was a great admirer of Galileo, a friend, and a fellow Tuscan. He had supported him in a previous scientific dispute in 1611 and employed as his personal secretary one of Galileo's closest acquaintances. Barberini's opposition to the recommendation of the consulting panel was "instrumental in preventing the direct condemnation of Galileo and the formal condemnation of Copernicanism as a heresy."[154] He believed that the physical truth of the Copernican system was a scientifically open question that should remain theologically open as well. "Though there were advocates both for and against the Copernican theory, the Church was not prepared to pass definitive judgment."[155] As a result the Inquisition chose to leave the truth status of Copernicanism open to various interpretations and they temporarily suspended circulation of Copernicus's book "until corrected." The "corrections" amounted to very little beyond editing "or amending the rare passages where the motion of the Earth was clearly acknowledged" as physically true and the recommended edits were issued as a separate page that was to be interpolated into any revised editions of Copernicus' work.[156] However, Copernicus' book

152 Drake, *Galileo*, 74.

153 Shea and Artigas, *Galileo in Rome*, 86.

154 Maurice A. Finocchiaro, *The Galileo Affair: A Documentary History* (Berkeley: University of California Press, 1989), 32.

155 David M. Miller, "The Thirty Years War and the Galileo Affair," *History of Science* 46 (March 2008), 49–74, 53.

156 Shea and Artigas, *Galileo in Rome*, 87.

remained in circulation and since no publisher was interested in printing the corrections, a revised edition never appeared. Libraries possessing copies of Copernicus' original work were expected to write in the corrections by hand and this was done for about half the copies in Italy. Throughout the rest of Europe virtually no one bothered to make the corrections.[157]

Before the decree mandating the corrections to Copernicus' book was published Cardinal Bellarmine met with Galileo privately. Bellarmine had said previously "that there was no thought of banning Copernicus's book, but at most of removing some passages in it and leaving his astronomical hypothesis intact."[158] To keep Galileo on the safe side of the ongoing debates within the Church Bellarmine advised him "not to hold or defend" the Copernican theory as physically proven. This meant that Copernicus' theory could still be discussed as a hypothesis. To speak "suppositionallly" about the Earth's movement was quite different from "holding" or "defending" the theory by offering what one held to be a "true demonstration" that the sun is at the center of the cosmos and that the Earth moved around it. "Treating Copernicanism as a mathematical construction was quite acceptable as long as he did not assert that it represented reality." This meant that he was welcome "to argue about its virtues as a model … especially if he was comparing it to the other models available."[159] To "hold" or "defend" a proposition was to say that the proposition had been clearly proven or demonstrated. "To weigh all the arguments on both sides of an issue was not regarded as defending one side against the other."[160] Bellarmine "kept abreast of scientific developments"[161] and was willing to consider the scientific truth of the Copernican theory if it could be proven or demonstrated.[162] "If there were such a demonstration, Bellarmine admitted the situation would be quite different. But to his knowledge no demonstration of this type existed, and he had grave doubts about whether it ever could."[163] Bellarmine also said that if such a demonstration of the Earth's motion could be found then certain Bible verses would need to be reinterpreted in light of the scientific discovery. The interpretation of Scripture however was

157 Shea and Artigas, *Galileo in Rome*, 87.

158 Drake, *Galileo*, 74.

159 Hannam, *God's Philosophers*, 325.

160 Drake, *Galileo*, 80.

161 Shea and Artigas, *Galileo in Rome*, 35.

162 McMullin, "Galileo on Science and Scripture," in *Cambridge Companion to Galileo* (Cambridge: Cambridge University Press, 1998).

163 William A. Wallace, "Galileo's Trial and the Proof of the Earth's Motion," *Catholic Dossier* 1.2 (1996) 7–13, 9.

not the business of Natural Philosophers or astronomers like Galileo, but of professional theologians.[164] "Church teaching would not be changed lightly; certainly not in the face of an unproven scientific theory." The Bible was not to be reinterpreted according to every theoretical whim of science. "Assertions that Copernicus's theory was *true*, or that exegetical principles should be adjusted to accommodate the theory, were not to be permitted in the absence of convincing proof."[165]

Galileo agreed to Bellarmine's request and thus there was no need for a formal legal injunction.[166] As he wrote to the Tuscan Secretary of State, Curzio Picchena, Galileo was "not overly concerned." The Church, he wrote, "had not definitively condemned Copernicanism, thus rebuking those calling it (and him) heretical. The decree did not mention him personally, nor was he deeply involved in the matter."[167] Yet, knowing that he had made a number of enemies (among both scientists and philosophers) who were keen on humiliating him, Galileo asked Bellarmine to give him a certificate stating what he had asked Galileo to agree to. The certificate stated "that he had been merely informed of the Index's decree and not personally condemned." Galileo also asked Pope Paul v for protection. During a private audience with the Pope (a meeting which in itself was a great honor) Galileo lamented "the malice of his persecutors." The Pope put him at ease telling him that he and the Inquisition would protect him from his enemies. Describing this meeting in a letter, Galileo writes, "when I showed signs of being still somewhat anxious about the future because of the fear of being pursued with implacable hate by my enemies, he cheered me up and said that I could put all care away because I was held in such esteem by himself and the whole Congregation that they would not lightly lend their ears to calumnious reports, and that I could feel safe as long as he was alive. Before I left he assured me several times that he bore me the

164 In 1615 Bellarmine's friendly advice to Galileo was that he should not venture outside mathematics and physics and he should avoid provoking theologians by teaching them how to correctly read the Bible.

165 Miller, "The Thirty Years War and the Galileo Affair," 53.

166 We know from the files of the Inquisition that an official injunction asking Galileo to not "hold, teach or defend" Copernicus' theory in "any way whatsoever," was written up but never signed and never served. The Inquisition Minutes, from the Archives of the Holy Office, tell us what actually happened on February 25th and 26th. There is no mention in it of Galileo's being instructed not "to hold, teach, or defend Copernicanism in any way whatever."

167 Miller, "The Thirty Years War and the Galileo Affair," 52.

greatest goodwill and was ready to show his affection and favor towards me at all times."[168]

In the succeeding years Galileo put the issue of the Earth's motion on the back burner and worked on many other important questions in physics. He continued to enjoy the welcome of high ranking church officials and the Roman elite and received letters of recommendation from Cardinals Alessandro Orsini and Francesco Maria del Monte. In 1620, Cardinal Maffeo Barberini even penned a poem in his honor where he celebrated Galileo's celestial discoveries. The letter in which Cardinal Barberini sent Galileo the poem was signed "your brother," a gesture of unusual warmth and affection.[169] Three years later Cardinal Barberini was elected Pope Urban VIII and Galileo honored him with the dedication of his new book *The Assayer*, which was not only an explicit discussion of comets and scientific method and an explicit critique of the Tychonic system, but also an implicit defense of the Copernican system.[170] Urban appreciated the gesture and was delighted with the content of book. Galileo was emboldened by the Pope's response and was "encouraged by the Pope's personal views regarding Copernicanism."[171] Urban had asserted that the Church had not condemned Copernicanism, nor was it about to condemn it as heretical.[172] It was simply premature to declare the physical truth of the theory. "Urban was inclined to entertain the possibility that the Earth revolved around the Sun," but "he was sceptical the matter could be proved either way."[173]

Urban's enthusiasm for *The Assayer* encouraged Galileo to pursue the Copernican research program and in 1624 he travelled to Rome to speak with the Pope about his proposed project. Galileo stayed about six weeks and "was warmly received by Church officials in general and the pope in particular, who granted him weekly audiences."[174] While Galileo and the Pope strolled around the gardens of the Lateran Palace, Galileo explained his proposed demonstration of the earth's motion from the ebb and flow of the tides. Urban was not impressed with the tidal argument and discouraged Galileo from using

168 Shea and Artigas, *Galileo in Rome*, 90.

169 Shea and Artigas, *Galileo in Rome*, 97.

170 Finocchiaro, *The Galileo Affair*, 32.

171 Miller, "The Thirty Years War and the Galileo Affair," 57.

172 "Urban VIII did not think Copernicanism to be a heresy or to have been declared a heresy by the Church in 1616." Finocchiaro, *The Galileo Affair*, 32.

173 Miller, "The Thirty Years War and the Galileo Affair," 57.

174 Finocchiaro, *The Galileo Affair*, 32.

it.[175] There could be other causes for the tides—such as the force that Kepler had suggested in his lunar attraction theory of the tides. Urban was a skeptic with regard to whether astronomy and mathematics could ever give us absolute certainty and he explained to Galileo that "it was beyond man's ability to work out how the heavens really worked. Whatever mathematical model was used to describe the movement of the planets, God could easily have arranged things differently but so as to give the same result. Thus, no matter how exactly any model predicts the movement of the planets, we cannot be sure that it is true."[176] To construct "a model that would make accurate astronomical predictions," Urban explained, "was not to prove the truth of that model."[177] Urban nevertheless granted Galileo the freedom to write his proposed book—as long as he treated the Copernican system as a hypothesis and not as a proven fact. Pope Urban also asked Galileo to include his "methodological argument that a natural phenomenon (like the tides) might have several possible causes, some of which may be unknowable, and so we cannot assign it a single cause with absolute certainty."[178]

Buoyed by his conversations with the Pope, Galileo left Rome and in 1624 started writing his *Dialogue on the Tides* which, at the recommendation of Urban, was eventually given the title *Dialogue Concerning the Two Chief World Systems of the World—Ptolemaic and Copernican*.[179] Completing his *Dialogue* in 1629, Galileo went to obtain an official license (or *imprimatur*) to print it in 1630. Neither of the censors who read Galileo's text "raised any serious objection, merely small points that Galileo could correct before coming back to Rome for final approval and printing of the text."[180] There was nothing heretical in Galileo's *Dialogue* and in the end it received the Church's imprimatur after scrutiny by censors in Florence and Rome.[181] While in Rome Galileo enjoyed a friendly meeting with the pope and his nephew Cardinal Francisco Barberini. He had a long private interview with Urban, and though the precise details of their conversation are unknown it is clear that Urban felt that he and Galileo had reached a mutual understanding. "A demonstration by Galileo of the superiority of the Copernican hypothesis to its competitors" would be

175 Wallace, "Galileo's Trial," 10.

176 Hannam, *God's Philosophers*, 326.

177 Lindberg, "Galileo, The Church and the Cosmos," 51.

178 Principe, *Scientific Revolution*, 62.

179 Stillman Drake, *Essays on Galileo and the History and Philosophy of Science*, ed. Noel M. Swerdlow and Trevor Harvey Levere (Toronto: University of Toronto Press, 1999), 163.

180 Heilbron, *Galileo*, 299.

181 David Miller, "The Thirty Years War and the Galileo Affair," 62.

"useful to the church" as it demonstrated to the Protestants "that the Roman Catholic Church had understood the astronomical and physical arguments." When Galileo left to return to Florence he felt that he had "completed his business to his entire satisfaction." At this time a contemporary described Galileo as "esteemed and honored by the entire [Papal] Court, as he deserves to be." Indeed, Galileo was virtually a member of the Papal Court and he was featured "on an official list of savants whose presence in Rome between 1630 and 1633 distinguished Urban's reign."[182]

In his *Dialogue* Galileo sought to demonstrate unmistakably that although Copernican heliocentrism was a hypothesis (as he and Urban had discussed), it was the *best* hypothesis available. "As the book emerged from his pen, however, it had become nothing less than a powerful argument on behalf of the indubitable *truth* of heliocentrism" and "no reader could have understood it otherwise." Galileo's intentions were clear and one did not have to read between the lines to perceive his purpose, "for in the *Dialogue* itself he repeatedly claimed to have demonstrated the 'truth' of his conclusions."[183] Galileo's *Dialogue* quickly dismisses the Ptolemaic model as untenable—which was a relatively uncontroversial move that simply affirmed the scientific consensus of the time. The unequivocal elimination of the Ptolemaic possibility would, in the minds of his contemporaries, narrow the choice down to either Copernicus or Tycho. However, Galileo "never mentions the Tychonic system by name, continues to refer to the alternative to Copernicus as Ptolemy, and never says a word about how a Copernican would defend Copernicus against Tycho."[184] The fact "that every astronomer was familiar with a third system (the Tychonic compromise) is mentioned nowhere in the book."[185] Nor does Galileo give any acknowledgment of the fact that "Ptolemy's system had been abandoned by all leading astronomers, whether Copernican or not." Galileo treats the basic assumptions of Tycho's system as non-controversial. "That at least the non-terrestrial planets go round the Sun is described as 'certain' , 'true', 'indubitable,' and 'the true arrangement; of which we are 'certain.'"[186] His explicit text "argues against a system which no knowledgeable anti-Copernican any longer defended" but he implicitly critiques the Tychonic system by describing how complex a model with two centers would have to be in order to account of the actual motions of the planets. "Galileo stops short of a direct confrontation with what (given the

182 Heilbron, *Galileo*, 299.
183 Lindberg, "Galileo, The Church and the Cosmos," 51.
184 Margolis, "Tycho's system and Galileo's *Dialogue*," 259.
185 Margolis, "Tycho's system and Galileo's *Dialogue*," 261.
186 Margolis, "Tycho's system and Galileo's *Dialogue*," 264.

role of the Collegio Romano) had effectively become the official astronomy of the Church" but he nevertheless sows the seeds of a debate that could not fail to materialize once the published book began to be discussed.

Demolishing a Ptolemaic system that none of his contemporaries ascribed to, Galileo champions a Copernican system that was not that of Copernicus (or even Kepler). What Galileo defends as demonstrated truth in his *Dialogue* "was not Copernicus' theory per se" but rather "a physical realization of parts of Copernicus' theory, which, by the way, dispensed with all the mathematical trappings (eccentrics, epicycles, Tusi couples and the like)."[187] As an appeal to a popular audience he presents a greatly simplified, non-technical, and highly accessible version of the Copernican system, which, if tested by astronomers, would have failed to yield accurate predictions of celestial phenomena. "In presenting the Copernican system to readers of the *Dialogue*, Galileo not only ignored the elliptical planetary orbits that had been introduced by Kepler, but oversimplified the astronomy even of Copernicus by treating the sun as if it were at the centre of all planetary orbits."[188]

The reason that Galileo presents such an oversimplified version of Copernicus is because he really wanted to talk about the mechanics of motion as it related to the Earth's movement as the cause of the tides. Presenting his theory of the tides, Galileo respectfully rejects Kepler's competing theory as pseudo-science. In the *Dialogue* Galileo writes of Kepler as "among the great men who have philosophized about this remarkable effect" of the tides and he goes on to describe Kepler as "a man of open and acute mind."[189] However, it is precisely for these reasons that Galileo expresses utter astonishment at Kepler's lunar origin explanation of the tides. Galileo "could not believe that the tremendous amount of water being raised and lowered was the result of an invisible force coming from the Moon, and furthermore this mysterious unseen force could pick up millions of tons of water without disturbing so much as a grain of sand on the beach."[190] Galileo dismisses Kepler's theory of the tides as sheer mysticism that relies on miraculous and mysterious forces. "To Galileo's mind, the hypothesis of an attraction of the kind suggested by Kepler

187 Peter Machamer, "Galileo Galilei," *The Stanford Encyclopedia of Philosophy* (Winter 2014 Edition), ed. Edward N. Zalta, URL = https://plato.stanford.edu/archives/win2014/entries/galileo/.

188 Stillman Drake, "Kepler and Galileo." In *Kepler: Four Hundred Years*, ed. A. Beer and P. Beer (New York: Pergamon Press, 1975), 237–47, 245.

189 Drake, "Kepler and Galileo." In *Kepler: Four Hundred Years*, 246.

190 Rocky Kolb, *Blind Watchers of the Sky: The People and Ideas that Shaped Our View of the Universe* (Oxford: Oxford University Press, 1999), 94.

was equivalent to the invoking of an occult quality."[191] For any explanation of the tides to be truly scientific, argues Galileo, it "must involve motion of the earth."[192] He says, "short of invoking miracles … there is no way to explain great and recurring motions of large seas on an absolutely fixed earth."[193]

Galileo knew well "how strongly, even emotionally, Urban clung to his almost unanswerable argument" that God could have devised other physical causes for the tides besides the movement of the Earth. It is likely that Urban wanted Galileo to put his argument in the narrator's own voice at the end as a mediating statement and final though regarding the epistemic limits of human knowledge in the scientific search for truth. Instead, Urban's argument was placed on the very last page of the *Dialogue* in the mouth of Simplcio—a hopelessly ill-equipped simpleton who is made to play the fool throughout the book as Salviati and Sagredo, the other two dialogue partners (who represent Galileo's own views), patronize and insult him.[194]

Needless to say, Urban was not pleased when reports of the published book reached his ears. At the same time, Urban was informed of the 1616 agreement between Galileo and Cardinal Bellarmine which Galileo had neglected to tell Urban about. Urban was also shown the legal injunction of 1616 (which should have been destroyed—as it was unsigned, never served, and had no legal value). "The Pope had no reason to disbelieve it, and since Galileo had never told him of any personal injunction not even to discuss Copernicus, it appeared to Urban that a legal order had been disobeyed." We don't know who dug out this injunction from the files, but the most likely candidate is the Jesuit astronomer Christopher Scheiner who was a onetime ally of Galileo who had been made into his enemy through a process of initial indifference and then extended ridicule.[195] Regardless of who showed the Pope the injunction,

191 Drake, "Kepler and Galileo." In *Kepler: Four Hundred Years*, 241.

192 Drake, *Galileo*, 93.

193 Drake, *Galileo*, 91; "Kepler did not live to read Galileo's theory, published in the Dialogue, but long before this time he had disparaged the idea that terrestrial motion could affect the waters." Stillman Drake, "Kepler and Galileo" in *Essays on Galileo and the History and Philosophy of Science*, ed. Noel M. Swerdlow and Trevor Harvey Levere (Toronto: University of Toronto Press, 1999), 343.

194 Simplcio was "a gentle idiot, stuffed with the learning of the schools, unable to think for himself, as immovable as the Aristotelian earth, in short, a comic caricature of all the philosophers who had had the effrontery to dismiss the revelations of Galileo." Heilbron *Galileo*, 230.

195 Originally "Scheiner had regarded Galileo as a comrade in arms, sometimes rash but often right." He supported Galileo's discovery of lunar mountains in 1615 and Galileo's proof of Venus's circumsolar orbit. He defended Galileo's priority over Simon Mayr in the

the document found in the Inquisition records "convinced the Pope that Galileo had deliberately deceived him." Having been deceived, humiliated, used, taken for granted, and personally betrayed by Galileo, Urban became furious. Moreover, Galileo's personal betrayal of Urban's friendship came at the worst possible moment. When Galileo's *Dialogue* arrived hot off the presses in 1632, Urban was overwhelmed by diplomatic negotiations and mounting criticism regarding the on-going Thirty Years War (where he was being blamed for the recent Protestant victories that placed the armies of the Reformation just across the Alps from Northern Italy). To make matters worse, the Hapsburgs spread the rumor "that the Pope was secretly in league with the Protestants, or even a Protestant himself, and threatened to have him deposed by convening a general Council." On the home front Urban was being "criticized for his nepotistic policies." Moreover, astrologers throughout Europe had been predicting the imminent death of the Pope, and by 1628 rumors of his death fueled by Spanish propaganda led the Spanish to call for a new papal conclave as if the papal seat was already vacant. In short, "Urban had weightier matters than Copernicanism in mind." To regain his dwindling political assets, Urban needed to show Europe that he was alive and well and had his house in order. Being "accused of Protestant sympathies himself, he certainly could not appear to shelter a possible heretic." Given his political situation Urban "needed to make an example of someone"[196] and at the same time "Galileo's bitter debates with some Jesuit astronomers made them into his enemies and pushed them into seeking vengeance by convincing the pope to bring Galileo to trial."[197]

discovery of Jupiter's satellites. He presented Galileo with several of his books dealing with telescopes, vision, the shape of the sun, and the flaws of heliocentrism as elaborated by Johannes Locher. "But his efforts to open a correspondence with 'the distinguished and brilliant Italian mathematician ... in the hope that from it greater light be shed on the truth' met with the same silence as his presents and concessions." After long ignoring him, Galileo later insinuated publicly that Scheiner had plagiarized Galileo's work on sunspots. Galileo dismissed offhand one of Scheiner's better discoveries as trivial and mistaken and Scheiner was also shamelessly attached and ridiculed in Galileo's *Dialogue*. Scheiner believed that Galileo's argument in the third day of the *Dialogue* (which in fact had been licensed before Galileo saw Scheiner's book), was based on information obtained from his book and applied to support Copernicus. "Having moved to Rome in 1624, Scheiner was in a position to stir up the Inquisition." Drake, *Galileo*, 94.

196 Miller, "The Thirty Years War and the Galileo Affair," 62.

197 Feldhay, *Galileo and the Church*, 17; "The reconstruction of the dialogue between Galileo and the Jesuits accounts for the danger the Jesuits were exposed to with the publication of the Dialogue Concerning the Two Chief World Systems (1632). Not only did it disclose the affinity between Galileo's ideas and the Jesuits', but it also pushed these ideas beyond

Political expediency met with scientific rivalry and Galileo was summoned to stand trial in Rome for his violation of the 1616 injunction. It was Galileo's alleged violation of a legally binding contract that was to be examined at the trial—not the status of Copernicanism as a philosophical system or scientific theory. Indeed, "No scientific question was raised at the trial."[198]

When Galileo arrived in Rome in February 1633 he lodged with "a new and very friendly Tuscan ambassador" who knew enough about the impending charges to tell Galileo "that all that was really in question was the 1616 meeting at Bellarmine's residence." At his trial Galileo "admitted receiving from Bellarmine the warning that the earth's motion could not be held or defended, but only discussed hypothetically." He avowed that he never received a special injunction not to discuss the Copernicanism "in any way whatever," and in his defense he produced the certificate that he had obtained from Bellarmine in 1616, wherein he was asked and agreed that he would not hold or defend Copernicanism as a proven fact. Galileo also claimed that in writing his *Dialogue*, he did not intend to actually defend the earth's motion as true. The inquisitors accepted Galileo's claim about his intentions and having been surprised by Bellarmine's certificate they called a recess for three weeks to decide on the next step. Because no signed document was ever found to support the memorandum on which the Inquisition had based its charge, by the rule of best evidence, Galileo had won on the only substantive issue raised. Instead of pursuing the case in a full trial, the Father Commissary of the Inquisition approached Galileo to offer an extrajudicial settlement, having sought and gained approval from Galileo's friend and the pope's nephew Cardinal Francesco Barberini. The inquisitors offered Galileo an out-of-court plea bargain: they would not press the most serious but least provable charge (violation of the special injunction), if Galileo would plead guilty to a lesser but more provable charge (transgression of the warning not to defend Copernicanism as a proven fact). Galileo accepted the deal and "admitted that he had erred by overstating the case for Copernicanism, but without any malicious intent."[199] Under normal circumstances that would have been the end of the trial, and at that point Galileo would have been sent home—having received a mere "slap on the wrist."

---

the limits the Jesuits resolved to sustain. Finding themselves under pressure from two extremes—the Dominicans who accused them of dissent from orthodoxy on the one hand and Galileanism which strove to push them toward the full consequences of their own intellectual tendencies on the other—the Jesuits were forced to retreat from backing Galileanism."

198 Drake, *Galileo*, 96.

199 Miller, "The Thirty Years War and the Galileo Affair," 64.

Such a mundane outcome to the trial would not serve the Pope's larger political purposes, however, and at that point—after the plea bargain had been arranged by the Inquisition and accepted by Galileo—"the Pope ordered special, extra-legal treatment for Galileo." Galileo "had become a pawn in a political game" and Urban needed to publicly make an example out of him if he "wanted the Catholic world to know that he led the fight against heresy in all of its forms, both within Italy and abroad."[200] It seems clear from what followed that Urban did not want any *actual* harm to come to Galileo, but that he wanted the larger public to *believe* that Galileo was being dealt with harshly. To undertake "this careful handling of appearances" Urban appointed his loyal nephew (and Galileo's friend) Francesco Barberini as the head of the commission. In this way "Urban could maintain firm, personal control of the case."[201] Desirous to convey "the appearance of a zealous and hard-line defender of the One True Faith" who would not tolerate any flirtations with Protestantism, "Urban crafted a public appearance of severity" while privately treating Galileo in an "unprecedentedly benign" manner. As historian David Miller explains, "the rhetoric surrounding Galileo was always harsher than his actual treatment."[202] For instance, a letter from Tuscan ambassador, Francesco Niccolini written during the proceedings of Galileo's trial reveals how Galileo's treatment was anything but ordinary. Niccolini writes:

> … I can add that on Tuesday morning he [Galileo] presented himself to the Father Commissary of the Holy Office, who received him in a friendly manner and had him lodged in the chambers of the prosecutor of that Tribunal, rather than in the cells usually given to criminals; thus, not only does he reside among the officials, but he is free to go out into the courtyard of that house…. Indeed, there is no precedent of anyone ever having been interrogated during a trial without being detained in a prison cell. Furthermore, they even allow his servant to wait on him, to sleep there, and, what is more, to come and go as he pleases, and they allow my own servants to bring him food to his room from here and to return to my house morning and evening.[203]

At the time of Galileo's trial, Copernicus' theory had not been declared a heresy. "The Church had no direct stake in which system was true; neither

200 Miller, "The Thirty Years War and the Galileo Affair," 65.
201 Miller, "The Thirty Years War and the Galileo Affair," 63.
202 Miller, "The Thirty Years War and the Galileo Affair," 64.
203 Finocchiaro, *Galileo Affair*, 250.

geocentrism nor Aristotelianism was ever Church dogma."[204] The "heresy" referenced in the charge against Galileo was the suspicion that Galileo had willfully disobeyed an official order given to him by Cardinal Bellarmine. For the charge of 'vehement suspicion of heresy' "it was sufficient to have disobeyed an official order." In the end, Galileo was found to be guilty of violating his agreement with Bellarmine and ordered to publicly abjure the Earth's motion, which he did. Yet, several cardinals, including Urban's own nephew Francesco Barberini, refused to sign the sentence against Galileo.[205] "Galileo was never—folklore aside—condemned as a heretic, imprisoned, or chained."[206] Galileo ended up under "house arrest" at his villa and estate in the Tuscan hills, where he would continue to work, train students, and write books.

In the midst of the troubling time of Galileo's trial in Rome Archbishop Ascanio Piccolomini of Siena invited Galileo to be his guest when the ordeal was over. During their time together the Archbishop displayed an exceptional "humanity and understanding" towards Galileo and "he managed to get Galileo's mind back on science, encouraging him to begin writing his long-planned treatise on motion."[207] Greatly encouraged by the Archbishop, "Galileo went back to his earlier research on motion, organized his notes, and five years later published his most important contribution to physics, the *Two New Sciences* (1638)."[208]

The Galileo Affair: From Conflict to Creative Tension and Creative Mutual Interaction

Was the Galileo Affair a clear instance of conflict between science and religion? From the best available evidence this does not appear to be the case. "The relevant documents show that many churchmen were on his side and many scientists were critical of him."[209] Harrison believes "it might be better to characterize the episode as a conflict *within* science (or, more strictly,

---

204 Principe, *Scientific Revolution*, 61.

205 Drake comments "That three cardinals of the Inquisition should refuse to sign the sentence against Galileo seems odd—unless they were personally certain of his devout Catholicism." Drake, *Galileo*, xxii.

206 Principe, *Scientific Revolution*, 62.

207 Drake says the Archbishop's care for him "literally saved Galileo's life and sanity." Drake, *Galileo*, 96.

208 Maurice A. Finocchiaro, *The Essential Galileo* (Indianapolis, IN: Hackett Publishing, 2008), 16; Finocchiaro adds "Without the tragedy of the trial, he might have never done it." *Essential Galileo*, 16.

209 Maurice A. Finocchiaro, "The Galileo Affair from John Milton to John Paul II: Problems and Prospects," *Science and Education* 8:2 (March 1999), 189–209, 194.

within astronomy and natural philosophy) rather than between science and religion."[210] Finocchiaro argues that "the real conflict was between" two different types of attitudes towards both science and religion—a conservative attitude and a progressive attitude. Finocchiaro points out that this tension is even more difficult to characterize because even "*within* the minds of many key players, including Galileo, Cardinal Bellarmine, and Pope Urban VIII" one finds "a split between conservative and innovative tendencies." Feldhay adds to this pointing out that the Catholic Church was not a monolithic institution at this time "in its relation to scientific knowledge in general and Copernican astronomy in particular" and "the different cultural orientations of groups within the church had dictated different attitudes towards Galilean science"—the Dominicans representing its conservative wing and the Jesuits its progressive wing.[211] In his philosophy and practice of science Galileo was much more of a Jesuit than a Dominican, but later Galileo alienated himself from many of his Jesuit supporters on account of his personal ambition and his unwillingness to share the spotlight, and his refusal to discuss his own scientific interpretations with his peers in a collegial manner.

It is likewise not obvious that the cautious approach to science voiced by many of Galileo's critics was entirely without warrant or justification. Galileo had not clearly demonstrated the motion of the Earth and his favorite proof—his argument from the tides—was empirically mistaken. Furthermore, even many of his Copernican peers, such as Kepler, believed that the argument for the Earth's motion from the tides was wrong. Historian of Science Thomas Dixon thus holds that at the time of Galileo's trial "an objective observer would have pronounced the scientific question an open one." The motion of the Earth would not be convincingly demonstrated for another two hundred years. As Wallace points out, "It is generally agreed that the earth's motion was not completely accepted until Friedrich Bessel's measurement of stellar parallax in 1838 and Leon Foucault's experiments with the pendulum in 1851—both still a long way off in 1633."[212]

From another angle one can view "the dialectic of conservation and innovation" as a creative tension that worked to fuel the progress of science. Owing to this creative tension, Finocchiaro documents a significant "change or development in Galileo's epistemological and methodological view of the nature of science." Galileo "certainly began with a version of the Aristotelian ideal of science as demonstration", but "by the end of his life, and certainly in the mature

210 Harrison, *Territories*, 173.

211 Feldhay, *Galileo and the Church*, 68.

212 Wallace, "Galileo's Trial," 12.

science of the Dialogue, Galileo held a fallibilist, probabilist, and hypothetical epistemology." This change in Galileo's epistemology is both "a historically real development" and "a progressive development, a change for the better." And the reason for this development—which pushed Galileo towards a more modern conception of science—was, says Finocchiaro, "pressure from the Church." That is, he says "the various conservative and reactionary elements which opposed, criticized, and even persecuted Galileo" performed "a valuable service, by making him see the light, as it were, in epistemological matters." The arguments of Galileo's critics ultimately "helped him to understand that the Copernican hypothesis had still not been conclusively demonstrated; this perception suggested that the search for a conclusive demonstration is an essential stage of scientific investigation." Thus even where there is a clear conflict in the Galileo Affair, it happens to be a type of conflict "in which religion had a beneficial influence on science."[213]

Conclusion

In his best-selling book *Atheist Universe*—hailed by Richard Dawkins as "an admirable work"—David Mills claims that "the Christian Church fought bitterly throughout its history—and is still fighting today—to impede scientific progress." To support this claim Mills continues, "Galileo, remember, was nearly put to death by the Church for constructing his telescope and discovering the moons of Jupiter. For centuries, moreover, the Church forbade the dissection of a human cadaver, calling it 'a desecration of the temple of the Holy Ghost.' Medical research was thereby stalled for almost a thousand years."[214] In part one of this essay we have seen why Mills' claims about both human dissection and Galileo are patently false. Yet, claims such as those found Mills' book abound, and war stories about science and religion are routinely employed in popular science writing and by the media. Mills' book is one of many contemporary expressions of what historians of science have called the Draper-White Thesis. Through entertaining storytelling, and a successful PR campaign, "Draper and White successfully transformed the age long *philosophical* tension between conservative tendencies versus innovative/speculative tendencies into a conflict between religion and science." While the warfare thesis is polemically attractive at a popular level, professional historians of science have shown that it is "historically bankrupt."[215] Instead of finding a state of historical warfare

---

213 Finocchiaro, "Science, Religion, and the Historiography of the Galileo Affair," 123.

214 David Mills, *Atheist Universe: The Thinking Person's Answer to Christian Fundamentalism* (Berkeley, CA.: Ulysses Press, 2006), 48.

215 Numbers, "Science and Religion," 80.

between science and religion there is a "growing recognition among historians of science that the relationship of religion and science has been much more positive than is sometimes thought."[216] The historical influences of the Christian faith upon the development of the natural sciences are numerous and can be found at the levels of "presupposition, sanction, motive, prescription, and substantive source."[217] Numerous studies in the history of science "have shown that Christianity has often nurtured and encouraged scientific endeavor, while at other times the two have co-existed without either tension or attempts at harmonization,"[218] and there are many cases where theistic presuppositions have played a constructive role in the shaping of scientific theories.[219]

Part 2: Theology and the Philosophical Foundations and Boundaries of Science

Emerging from the philosophy of the 18th century European Enlightenment, the fundamental distinction between the *realm of fact* and the *realm of value* continues to hold sway at the beginning of the 21st century. It is often said that "science deals only in facts" and that "values come in only when decisions are made as to how the facts of science are to be applied."[220] Yet, from physics to philosophy to theology, there has been a "quiet revolution" taking place that has seriously undermined the practice of drawing a sharp line between the objective and subjective realms of human experience.[221] As neuroscientists meticulously map the mind, and quantum physicists discover the role of the observer intricately inscribed into the basic equations of the cosmos, there is an ever increasing necessity to factor the phenomenon of the subjective self into the equations aimed at describing objective reality. In the 20th century the practice of science gave rise to a number of boundary questions regarding both the physical foundations of reality and the cosmic origins of existence.

216 Gary Ferngren, "Introduction," in *Science and Religion: A Historical Introduction*, ed. Gary Ferngren (Baltimore: Johns-Hopkins University Press, 2002), ix.

217 Finocchiaro, "Science, Religion, and the Historiography of the Galileo Affair," 115.

218 Ferngren, *Science and Religion*, ix.

219 Brooke, "Preface," ix.

220 John Dupre, "Fact and Value," in *Value-free science?: Ideals and Illusions*, ed. Harold Kincaid, John Dupré, Alison Wylie (Oxford: Oxford University Press, 2007), 27. Dupre is describing a stereotype of science that he does not hold to be true.

221 See Ted Peters, "Theology and the Natural Sciences," *Modern Theologians: An introduction to Christian theology in the twentieth century*, ed. David F. Ford, 2nd ed. (Cambridge, MA: Blackwell, 1997), 649.

As quantum physicists discovered that atoms don't exist when they're not being observed, and cosmologists became accustomed to speaking about the creation of something out of nothing, a horizon of mystery ironically began to emerge within the disciplines of the physical sciences that were the *best* understood.

At the same time, philosophers of science found that the practice of science cannot be neatly separated from its social and cultural context. As philosophers of science elucidated the foundational and general presuppositions underlying the scientific endeavor, historians of science disclosed particular historically-contextualized presuppositions that functioned in key episodes of scientific theorizing and discovery. Many of these historians were surprised to discover the role that religion played in supplying many of the key philosophical presuppositions that undergirded the early modern scientific enterprise. Recent scholarship within the history and philosophy of science has shown that in both the past and the present, theology has provided the "foundational assumptions for certain key scientists and scientific discoveries."[222] Historians of science have identified numerous "ways in which religious beliefs have influenced science." These ways include "presuppositions underwriting science ... sanctions and motives for doing science ... principles for regulating scientific methodology and for selecting acceptable theories," and so on.[223] Scholars in this area have found that "theological beliefs can have both internal and external influences on the development of science."[224] In the second part of this essay I aim to broaden our understanding of the nature of the interaction of science and theology by exploring the role that metaphysical presuppositions and religious concepts play within the scientific process.[225] I will examine the role of religious and theological thought within both the general philosophical

222 Alan G. Padgett, "Science and Theology," in *The Encyclopedia of Christianity*, vol. 4, ed. Erwin Fahlbusch, et al. (Grand Rapids: Eerdmans/Brill, 2005), 873.

223 John Hedley Brooke, *Science and Religion: Some Historical Perspectives* (Cambridge: Cambridge University Press, 1991), 18–33; and Edward B. Davis, "Christianity and Early Modern Science: The Foster Thesis Reconsidered," in *Evangelicals and Science in Historical Perspective*, ed. David N. Livingstone, D.G. Hart, and Mark A. Noll (Oxford: Oxford University Press, 1999), 77.

224 Peter E. Hodgson, *Theology and Modern Physics* (Burlington, VT: Ashgate, 2005), 16. See also Peter E. Hodgson, "Presuppositions and Limits of Science," in *The Structure and Development of Science*, ed. G. Radnitzky and G. Andersson (Dordrecht: Reidel, 1979) 133–146.

225 Following Stephen J. Wykstra who says, "Our vision of the nature of science needs to be broadened if we are to account for the roles that metaphysical and religious believing play within the scientific process." Stephen J. Wykstra, "Religious Beliefs, Metaphysical Beliefs,

presuppositions that undergird the whole scientific enterprise and within particular presuppositions that were present during pivotal episodes of scientific discovery. I will then discuss how particular social and cultural values still play an important role as shaping principles within science. The fact that science depends upon philosophy (which in turn borrows heavily from theology) means that science has limits. While its vital dependence upon philosophy limits the scope of science, boundaries also emerge within the context of scientific practice and as a result of particular scientific discoveries. Here I will discuss how two key theories from 20th century science—quantum mechanics and Big Bang Cosmology limit—in principle—the scope of future scientific exploration.

The General Presuppositions of Science

The scientific enterprise is founded upon a number of *general presuppositions* about the nature of reality. These presuppositions are non-empirical philosophical beliefs about things such as the orderliness and regularity of reality, the ontological *reality* of reality, the intelligibility and contingency of existent structures and entities, the agential passivity of non-conscious nature, the unity and uniformity of the physical universe, and so on. These presuppositions are *general* in that they necessarily precede and underpin *all* scientific experimentation and reasoning. And these general presuppositions are *a priori* "conditions that are necessary for the possibility of scientific activity as such, although they can be ignored by particular scientists."[226] As *preconditions* they are absolutely required for science to take place and are not open to experimental confirmation or falsification by scientific experimentation. The nature of these general presuppositions is such that "for science to develop, these beliefs must be held, at least implicitly, by society as a whole and by scientists themselves."[227]

Aesthetic, Epistemic, and Moral Values That Shape Science

1 The Aesthetic Value of Simplicity

The belief that simple theories are better than more complex theories is a foundational aesthetic value that guides the practice of science and it is one

---

and Historiography of Science," *Osiris*, Vol. 16, *Science in Theistic Contexts: Cognitive Dimensions* (2001) 29–46, 29.

226 Mariano Artigas, "Three Levels of Interaction Between Science and Philosophy," *Intelligibility in Science*, ed C. Dilworth (Amsterdam: Rodopi, 1992), 123.

227 Hodgson, *Theology and Modern Physics*, 16. See also Peter E. Hodgson, "Presuppositions and Limits of Science," 133–146.

of the most important philosophical assumptions undergirding a belief in the explanatory power of scientific reductionism. The idea of explanatory simplicity was first introduced by Aristotle (384–322 BCE) as a principle of parsimony which affirms that the simplest explanation for a given phenomenon is the one that will most likely be true. Aristotle states this notion as a fundamental assumption. He says, "We may assume the superiority *ceteris paribus* of the demonstration which derives from fewer postulates or hypotheses."[228] One should thus favor simpler theories and explanations over those that are more complex. Over a thousand years later, we find this philosophical principle affirmed in the theology of Thomas Aquinas (1225–1274): "If a thing can be done adequately by means of one, it is superfluous to do it by means of several; for we observe that nature does not employ two instruments where one suffices."[229] This approach to logic came to be called "Occam's Razor," after the logician and Franciscan friar William of Occam (1287–1347) who taught that explanatory entities should not be multiplied beyond necessity.

Emerging as a crucial logical axiom in the Christian theology of the Middle Ages Occam's Razor was an important guiding principle in shaping the foundation of early modern science. When Galileo compares the "Two Chief World Systems" that explain the motions of the planets (which at the time would have been that of Copernicus and that of Tycho Brahe) he assumes that there can be only *one* model of the solar system that is correct. This is because, Galileo explains, "Nature does not multiply things unnecessarily; that she makes use of the easiest and simplest means for producing her effects; that she does nothing in vain, and the like."[230] Later in the 17th century the well-known physicist Isaac Newton (1643–1727) includes Occam's Razor as one of his three "rules of reasoning in philosophy" in his *Principia Mathematica*: "Rule I: We are to admit no more causes of natural things than such as are both true and sufficient to explain their appearances."[231] Writing a few generations after Newton, the chemist Antoine Lavoisier (1743–1794) affirms that "It is, after all, a principle of logic not to multiply entities unnecessarily" and he applies this principle dutifully in his practice of science as he argues against hypothetical substances, such as phlogiston, as gratuitous suppositions. "If all of chemistry can be explained in

---

228 Aristotle, *Posterior Analytics*, transl. Richard McKeon (Oxford: Clarendon Press, 1963), 150.

229 Thomas Aquinas, *Basic Writings of St. Thomas Aquinas*, trans. A.C. Pegis (New York: Random House, 1945), 129.

230 Galileo Galilei, *Dialogue Concerning the Two Chief World Systems*, trans. Stillman Drake (Berkeley: University of California Press, 1962), 397.

231 Isaac Newton, *Principia Mathematica* (London, 1687), 41:1.

a satisfactory manner without the help of phlogiston," says Lavoisier, "that is enough to render it infinitely likely that the principle does not exist, that it is a hypothetical substance, a gratuitous supposition."[232] Writing more than 200 years later, Albert Einstein, agrees: "The grand aim of all science ... is to cover the greatest possible number of empirical facts by logical deductions from the smallest possible number of hypotheses or axioms."[233]

Today the philosophical centrality of Occam's Razor remains and "many scientists believe that simplicity is a crucial element in their quest for knowledge."[234] The vast majority of current practicing scientists believe that, all things being equal, simpler theories are better.[235] Philosopher of biology Elliot Sober explains that "scientists ... frequently appeal to parsimony to justify their choice of hypotheses" and that "removing the principle of parsimony from the organon of scientific method threatens to deprive science of its results."[236]

But why should scientists favor simpler theories over more complex ones? There is no *simple* answer to this question. "A problem with Occam's razor is that nearly everybody seems to accept it, but few are able to define its exact meaning and to make it operational in a non-arbitrary way."[237] There is no obvious logical or empirical connection between plausibility and parsimony. As philosopher of science Ernan McMullin says, "Efforts to express a criterion of 'simplicity' in purely formal terms continue to be made, but have not been especially successful."[238] Although the connection between simplicity and truth

---

232 Antoine-Laurent de Lavoisier, "Réflexions sur le Phlogistique," in *Oeuvres: Volume 2* (Paris: Imprimerie Impériale, 1862), 623–24.

233 Albert Einstein, quoted in Leonard Kollender Nash, *The Nature of the Natural Sciences* (Boston: Little, Brown, 1963), 173.

234 Hugo A. Keuzenkamp, Michael McAleer and Arnold Zellner, "The enigma of simplicity," in *Simplicity, Inference and Modelling: Keeping it Sophisticatedly Simple*, ed. Arnold Zellner, Hugo A. Keuzenkamp and Michael McAleer (Cambridge: Cambridge University Press, 2004), 1.

235 Alan Baker, "Simplicity," *The Stanford Encyclopedia of Philosophy* (Fall 2013 Edition), Edward N. Zalta, ed., URL = <https://plato.stanford.edu/archives/fall2013/entries/simplicity/>.

236 Elliott Sober, *From a Biological Point of View: Essays in Evolutionary Philosophy* (New York: Cambridge University Press, 1994), 140.

237 Hugo Keuzenkamp, Michael McAleer and Arnold Zellner, "The enigma of simplicity," in *Simplicity, Inference and Modelling: Keeping it Sophisticatedly Simple*, ed. Hugo Keuzenkamp, Michael McAleer and Arnold Zellner (Cambridge: Cambridge University Press, 2001), 1.

238 Ernan McMullin, "Values in Science," *PSA: Proceedings of the Biennial Meeting of the Philosophy of Science Association* 2 (1982), 3–28, 16.

is taken for granted by many practicing scientists, "There is no reason—in the absence of *independent* belief in the simplicity of nature," says philosopher of science James McAllister, "why that policy should result in hypotheses that are true more often than would any other."[239] Sober points out that it is "only because of a set of *background assumptions*" that parsimony is allowed to connect with plausibility within a particular research problem. However, says Sober, "what makes parsimony reasonable in one context ... may have nothing in common with why it matters in another. The philosopher's mistake is to think that there is a single global principle that spans diverse scientific subject matters."[240] In the end, it would seem that Occam's Razor is essentially an *aesthetic* value. Yet, *as* an *aesthetic* value it has played and continues to play a vital role in scientific explanation and theory choice.

2 The Aesthetic Value of Beauty

Another example of an aesthetic value within science is *beauty*. For the working scientist "beauty is thought (and felt) to lie in explaining much with little, and in finding pattern, especially simple pattern, in the midst of apparent complexity and disorder."[241] Within the physical sciences beauty is often held as a guide to truth. According to McAllister, "the history of science teems with instances in which indicators of beauty appear to have prevailed over empirical criteria in directing theory-formulation."[242] And in physics today, the appeal to beauty remains a central aspect of research motivation and theory choice. "Most of the great innovators in contemporary physics and cosmology have been strongly attracted by intellectual beauty and have combined this with faith that beauty will point the path to comprehension."[243] Historian of

239 James W. McAllister, "Truth and beauty in scientific reason," *Synthese* 78 (1989) 25–51, 32.

240 Sober, *From a Biological Point of View*, 140.

241 Herbert A. Simon, "Science seeks parsimony, not simplicity: searching for pattern in phenomena," in *Simplicity, Inference and Modelling: Keeping it Sophisticatedly Simple*, ed. Arnold Zellner, Hugo A. Keuzenkamp and Michael McAleer (Cambridge: Cambridge University Press, 2004), 33.

242 James W. McAllister, "Truth and beauty in scientific reason," *Synthese* 78 (1989) 25–51, 29. "In the history of science there exist many instances of theory-choice which cannot be explained without reference to these aesthetic criteria." McAllister, "Truth and beauty in scientific reason," 31.

243 Harold Osborne, "Mathematical Beauty and Physical Science," *British Journal of Aesthetics* 24 (1984), 291–300, 291; "Historically, this faith was actually vindicated to a great extent in the works of these scientists. Theories which they created on what were considered primarily aesthetic grounds were later confirmed experimentally." Gideon Engler, "Aesthetics in Science and in Art," *British Journal of Aesthetics* 30 (1990) 24–34, 24.

Science Thomas Kuhn points out that such mathematical beauty was so central to the Copernican astronomer Johannes Kepler that his "entire astronomical program is based in a metaphysical faith in mathematically expressed harmonies in nature."[244] A few centuries later, Einstein affirms that for physicists "the only physical theories that we are willing to accept are the beautiful ones."[245] Indeed, Einstein was resolutely skeptical of certain aspects of quantum physics because these parts of the theory were, in his assessment, not beautiful enough to be true. For instance, his "rejection of indeterminism was essentially aesthetic: for him the harmony of the universe would be marred if, to use his own metaphor, God cast dice."[246] Nobel Prize winning atomic physicist Paul Dirac became convinced of the truth of Einstein's general theory of relativity primarily because of the beauty of the theory: "One has a great confidence in the theory arising from its great beauty, quite independent of its detailed successes ... One has an overpowering belief that its foundations must be correct quite in dependent of its agreement with observation."[247]

Despite Einstein's resistance to embrace indeterminism, the founders of quantum physics were seeking beauty in their theorizing and others saw beauty in quantum theory. One of the founders of quantum physics, Werner Heisenberg once commented to Einstein: "I frankly admit that I am strongly attracted by the simplicity and beauty of the mathematical schemes which nature presents us."[248] Reflecting on Erwin Schrodinger's wave equation describing quantum phenomena, Dirac said: "It seems that if one is working from the point of view of getting beauty in one's equations, one is on a sure line of progress."[249] This is because, explains Dirac, "Schrodinger got this equation by pure thought, looking for some beautiful generalization of De Broglie's ideas, and not by keeping close to the experimental development of the subject in the way Heisenberg did."[250] Dirac even goes so far as to say that if one wishes

---

244 Roger Trigg, *Rationality and Science: Can Science Explain Everything?* (Oxford: Blackwell, 1993), 224.

245 Albert Einstein, quoted in Graham Farmelo, *It Must be Beautiful: Great Equations of Modern Science* (London: Granta Books, 2002), xii.

246 Mcallister, "Truth and beauty in scientific reason," 36.

247 James W. Mcallister, "Is Beauty a Sign of Truth in Scientific Theories?" *American Scientist* 86 (1998), 174–183, 174.

248 Werner Heisenberg, "Letter to Albert Einstein," in Ian Stewart, *Why Beauty is Truth: A History of Symmetry* (New York: Basic Books, 2008), 278.

249 Paul Dirac, "The Evolution of the Physicist's Picture of Nature," *Scientific American* 208:5 (1963), 45–53, 47.

250 Quoted in Mcallister, "Truth and beauty in scientific reason," 30.

to discover truth in physics, "it is more important to have beauty in one's equations than to have them fit experiment."[251]

The idea that beauty is a guide to truth remains important among current physicists as well.[252] "Contemporary science is infused with a powerful element of aesthetic faith ... It is a faith that aesthetically good theory will be confirmed by fact and experience because the universe itself is aesthetically structured."[253] For example, the physicist Steven Weinberg has recently reflected, "It is precisely in the application of pure mathematics to physics that the effectiveness of aesthetic judgments is most amazing.... Mathematical structures that confessedly are developed by mathematicians because they seek a sort of beauty are often found later to be extraordinarily valuable by physicists."[254] Weinberg explains that "time and again physicists have been guided by their sense of beauty not only in developing new theories but even in judging the validity of physical theories once they are developed. It seems that we are learning how to anticipate the beauty of nature at its most fundamental level. Nothing could be more encouraging that we are actually moving toward the discovery of nature's final laws."[255] The appeal to beauty is particularly prevalent among contemporary advocates of String Theory—a physical theory that is mathematically elegant but may never be empirically testable. Describing the early formulation of String Theory, John Schwarz reflects, "We felt strongly that string theory was too beautiful a mathematical structure to be completely irrelevant to nature."[256] Nobel Laureate and string theorist David Gross similarly remarks that "string theory could not be wrong because its beautiful mathematics could not be accidental."[257] Mathematical and theoretical physicist Edward Witten believes

251 Dirac, "The Evolution of the Physicist's Picture of Nature," 47.

252 Beauty is also important for non-physicists. James D. Watson reports that, when Rosalind Franklin learned of his and Francis Crick's model of the structure of DNA, she "accepted the fact that the structure was too pretty not to be true." J.D. Watson, *The Double Helix: A Personal Account of the Discovery of the Structure of DNA*, ed. G.S. Stent (London: Weidenfeld and Nicolson, 1968), 210; Evolutionary biologist Sean Carroll says "beauty, in science, is much more than skin-deep." Sean B. Carroll, *Endless Forms Most Beautiful: The New Science of Evo Devo* (New York: W.W. Norton & Company, 2006), 13.

253 Osborne, "Mathematical Beauty and Physical Science," 293.

254 Steven Weinberg, *Dreams of a Final Theory* (New York: Pantheon Books, 1992), 153.

255 Weinberg, 90.

256 John Schwarz, "Superstring-A Brief History," in *History of Original Ideas and Basic Discoveries in Particle Physics*, ed. H. Newman and T. Ypsilantis (New York: Plenum Press, 1996), 695–706, 698.

257 Leonard Susskind, "Quark Confinement," *The Rise of the Standard Model: Particle physics in the 1960s and 1970s* ed. Lillian Hoddeson, L. Brown, M. Riordan, and M. Dresden (Cambridge: Cambridge University Press, 1997) 233–243, 235.

that string theory must be true because of "its wonder, its incredible consistency, remarkable elegance and beauty."[258]

But why should physicists assume that beauty will point to truth? Although "much tribute has been paid" to the nature of beauty in the sciences, comments philosopher of aesthetics Harold Osborne, a "systematic analysis has not been attempted but … it is taken for granted that anyone with a talent for scientific matters will recognize a beautiful theory when he sees it."[259] There is no purely empirical reason or justification for affirming this aesthetic criterion and, as Nobel laureate Eugene Wigner remarks, the reason for the effectiveness of mathematical beauty in physics "is something bordering on the mysterious and there is no rational explanation for it."[260] While some are comfortable seeing the role of beauty within science as a mystery, others have asserted an explicitly *theological* justification for why physicists focus on beauty. Heisenberg says that expressions of beauty such as the "miracle of symmetry," harmony, and "the beauty of simplicity" reveal the "inner truth" of physical reality because they are reflections of "the original archetype of creation."[261] Dirac similarly affirms a divine origin for such beauty: "God used beautiful mathematics in creating the world."[262] Contemporary string theorists have likewise grounded the equating of truth and beauty within the Divine. Theoretical physicist Michio Kaku reflects that in string theory "the mind of God is music resonating through 11–dimensional hyperspace,"[263] and Harvard string theorist Lubos Motl comments that "Superstring/M-theory is the language in which God wrote the world."[264]

3 Epistemic and Ethical Values

In addition to aesthetic principles, which shape both the practice and content of science, there is also a central role for epistemic and ethical values within science. Since the 1960s historians and philosophers of science have increasingly recognized that science is *value-laden*—that values are an intrinsic

258 John Horgan, "Physics Titan Edward Witten Still Thinks String Theory Is 'On the Right Track,'" *Scientific American* (September 29, 2014) URL: https://blogs.scientificamerican.com/cross-check/physics-titan-still-thinks-string-theory-is-on-the-right-track/.

259 Osborne, "Mathematical Beauty and Physical Science," 292.

260 Eugene Wigner, "The Unreasonable Effectiveness of Mathematics in the Natural Sciences," *Communications in Pure and Applied Mathematics* 13:1 (February 1960): 1–14.

261 Engler, "Aesthetics in Science and in Art," 25.

262 John Polkinghorne, *The Particle Play* (New York: W.H. Freeman, 1979), 2 and 126.

263 Michio Kaku, "Interview on the Leonard Lopate Show," WNYC (January 2, 2004).

264 Quoted by Bert Schroer, "String theory, the crisis in particle physics and the ascent of metaphoric arguments," *International Journal of Modern Physics D Int. J. Mod. Phys. D*, 17, 2373 (14 Mar 2006) http://arxiv.org/abs/physics/0603112, 21.

component within scientific theorizing and scientific practice. As *assumptions* about the worth or goodness of something, values, as such, are not empirically testable. Even though popularizers of science have continued to promote the image of science as a value-free enterprise, philosophers of science have come to recognize that "value-free science is an unattainable or untenable ideal."[265]

McMullin explains that "there are certain characteristic *epistemic* values which are integral to the entire process of assessment in science."[266] The desire to have a value-free science is itself an epistemic value. *Objectivity* is likewise an epistemic value. "To call a thing objective implies that it has a certain importance to us and that we approve of it."[267] Another epistemic value, "namely truth itself, has always been recognized as permeating science."[268] In the classic account of science, says McMullin, truth was "the goal of the entire enterprise" and in the practice of science today "truth is still a sort of horizon-concept or ideal of the scientific enterprise, even though we may not be able to assert truth in a definitive manner."[269] In addition to objectivity and truth, McMullin lists the epistemic values of *unifying power* (a theory's ability to bring together previously disparate areas of inquiry), *external consistency* (a theory's consistency with other theories and with the general background of expectation), *internal coherence* (that there should be no logical inconsistencies and no unexplained coincidences within a theory), *fertility* (a theory's ability to make novel predictions that were not part of the set of original explananda) and *predictive accuracy*.[270]

There are also a number of moral values that guide and shape the workings of science. For instance honesty, openness, and integrity are "moral values which have always been seen as essential to the success of communal inquiry." Science is a communal enterprise that "cannot succeed unless results are

265 Heather Douglas, "Rejecting the Ideal of Value-Free Science," in *Value-free science?: Ideals and Illusions*, ed. Harold Kincaid, John Dupré, Alison Wylie (Oxford: Oxford University Press, 2007), 121.

266 McMullin, "Values in Science," 6.

267 Julian Reiss and Jan Sprenger, "Scientific Objectivity", *The Stanford Encyclopedia of Philosophy* (Summer 2016 Edition), Edward N. Zalta, ed., URL = <https://plato.stanford.edu/archives/sum2016/entries/scientific-objectivity/>.

268 McMullin, "Values in Science," 6.

269 McMullin, "Values in Science," 7.

270 McMullin, "Values in Science," 15–16; Thomas Kuhn similarly distinguished key epistemic values of the scientific enterprise such as accuracy, consistency, scope, simplicity, and fruitfulness (fecundity). See Thomas Kuhn. "Objectivity, Value Judgment, and Theory Choice." In *The Essential Tension: Selected Studies in Scientific Tradition and Change.* (Chicago: University of Chicago Press, 1977), 321–322.

honestly reported, unless every reasonable precaution be taken to avoid experimental error, unless evidence running counter to one's own view is fairly handled."[271] For science to make progress, scientists need to trust that the experimental results of other scientists are genuine and not falsified.[272] Moral principles shape science via ethical guidelines for conducting research on human and animal subjects, cultural norms and social values that determine the appropriateness of research topics (e.g., conservation biology, nuclear weapons research, genetic enhancement research), and the values of individual researchers.[273] Scientists' "personal beliefs may significantly influence the type of research problems that scientists may choose to work on, the approach that they use in addressing the problem, and the magnitude of effort and dedication that they invest in finding the solution to their chosen problem."[274] As chemist and philosopher Michael Polanyi pointed out, "Only a tiny fraction of all knowable facts are of interest to scientists, and scientific passion serves also as a guide in the assessment of what is of higher and what is of lesser interest ... This appreciation depends ultimately on a sense of intellectual beauty."[275]

The *practice* of science is likewise oriented towards outcomes that are ethically discerned and derived. The scientific endeavor to produce a vaccine is for the good of public health and for the good of humanity as a whole. The motivation behind investigating ecosystems is not typically for the sake of accumulating pure value-free knowledge, but rather in order to support efforts to conserve such ecosystems for the sake of the animals who live in them and the humans that enjoy them. "We are interested in scientific investigations that have consequences for action."[276] Science is pursued and funded according to

---

271 McMullin, "Values in Science," 7.

272 Hodgson, *Theology and Modern Physics*, 20. Regarding the justification of this presupposition see Meredith Wadman, "One in Three Scientists Confesses to Having Sinned," *Nature* 435, no. 7043 (June 9, 2005): 718–719. Wadman says that such "misconduct ranges from faking results outright to dropping suspect data points," 718–719.

273 Steven Shapin and Simon Schaffer, *Leviathan and the Air Pump* (Princeton: Princeton University Press, 1985). Other types of presuppositions enter into science as well. For example philosopher Michael Stenmark explains, "Scientific knowledge presupposes introspective knowledge and knowledge based on memory, then one first must know these things to be able to do science." Stenmark, *Scientism: Science, Ethics and Religion*, 119.

274 I.S. Caleon, G. Lopez Wui, and H.P. Regaya, "Personal Beliefs as Key Drivers in Identifying and Solving Seminal Problems: Lessons from Faraday, Maxwell, Kepler and Newton," *Science Education International* 26:1 (2015), 3–23.

275 Michael Polanyi, *Personal Knowledge: Towards a Post-Critical Philosophy* (London: Routledge, 1962), 143.

276 Dupre, "Fact and Value," 30.

the *relevance* of its findings in so far as they shed light on the things we most value. In this way, says Dupre, "fact and value are typically inextricably linked in the matters that concern us."[277]

The process of "scientific inference is regulated by normative rules" and these rules depend on diverse values. "Scientists try to construct *good* tests of their hypotheses, they judge some explanations *good* and others *bad*, and they say that some inferences are *flawed* or *weak* and others are *strong*." The italicized words in the previous sentence indicate "that scientists are immersed in tasks of evaluation. They impose their norms on the ideational entities they construct."[278] Values enter into the process of science at a number of levels—inspiration, motivation, theory construction, and theory justification. As Kuhn explains, "The criteria of [theory] choice function not as rules, which determine choice, but as values which influence it. Two men deeply committed to the same values may nevertheless, in particular situations, make different choices, as in fact they do."[279] Consequently, both epistemic and nonepistemic "values are logically needed for reasoning in science, even in the internal stages of the process."[280] Moreover, the presence of values within science is not a *bad* thing. In fact, science *should* have values. As philosopher of science Heather Douglas argues, a *value-free ideal* is a *bad ideal* for science. "In many areas of science, particularly areas used to inform public policy decisions, science should not be value free … In these areas of science, value-free science is neither an ideal nor an illusion. It is unacceptable science."[281]

General Metaphysical Presuppositions of Science

In addition to the moral values and aesthetic principles that shape science and guide scientific discovery and theory choice, there are also general metaphysical presuppositions that serve as the deeper philosophical foundations of the entire scientific enterprise. Metaphysical presuppositions that provide the necessary conditions for science include:

---

277 Dupre, "Fact and Value," 35.

278 Dupre, "Fact and Value," 110.

279 Thomas Kuhn, *The Essential Tension: Selected Studies in Scientific Tradition and Change* (Chicago: University of Chicago Press, 1977), 331.

280 Douglas, "Rejecting the Ideal of Value-Free Science," 121.

281 Douglas, "Rejecting the Ideal of Value-Free Science," 121.

1 A Belief That the Physical Universe is, in Some Sense Good, and Therefore Worthy of Careful Study

This first necessary condition for the existence of science affirms that one must consider the objects and goals of science as *valuable* and *worth* pursuing before one pursues the study and practice of science.[282] Biologist and Nobel Laureate Konrad Lorenz expresses the goodness and worth of physical reality in the language of love, and says that he and "all of the biologists [he] know[s] are undeniably lovers of their objects of study."[283] The presupposition of goodness or worth with regard to physical reality is often related to and conveyed by the appreciation aesthetic values such as awe, wonder, and beauty. Numerous scientists have thus affirmed that nature is *worth* studying because it is beautiful and because the study of nature fills one with awe. For instance, Heisenberg, reflecting on the process of scientific discovery in physics, says "What these internal relations show in all their mathematic abstraction, an incredible degree of simplicity, is a gift that we can only accept with humility. Not even Plato could have believed that it would be so beautiful."[284] Physicist Richard Feynman likewise expresses the goodness or worth of investigating physical reality through invoking the aesthetic values of awe and wonder: "The same thrill, the same awe and mystery, come again and again when we look at any problem deeply enough. With more knowledge comes deeper, more wonderful mystery, luring one on to penetrate deeper still. Never concerned that the answer may prove disappointing, but with pleasure and confidence we turn over each new stone to find unimagined strangeness leading on to more wonderful questions and mysteries."[285]

2 A Belief That the World is Orderly and Rational

If physical reality was assumed to be unstructured, disorderly, or fundamentally chaotic, science would be impossible.[286] The presupposition that order exists in nature is thus a necessary condition of scientific inquiry because if one did not believe that order existed at all in nature, then searching for it

---

282 Peter Hodgson, "Presuppositions and Limits of Science," 136.

283 Marco Bersanelli and Mario Gargantini, *Galileo to Gell-Mann: The Wonder that Inspired the Greatest Scientists of all Time* (Philadelphia: Templeton Press, 2009), 10.

284 Bersanelli and Gargantini, 6.

285 Quoted in Bersanelli and Gargantini, 7.

286 Trigg, *Rationality and Science*, 224.

scientifically would be pointless.[287] For example, Einstein's development of the general theory of relativity was premised on the assumption that the universe is a puzzle to be solved, and his lifelong search for a unified field theory (to unify general relativity with electromagnetism) assumed that there is a deeper cosmic rationality waiting to be discovered. As physicist Paul Davies comments, "All science proceeds on the assumption that nature is ordered in a rational and intelligible way. You couldn't be a scientist if you thought the universe was a meaningless jumble of odds and ends haphazardly juxtaposed. When physicists probe to a deeper level of subatomic structure, or astronomers extend the reach of their instruments, they expect to encounter additional elegant mathematical order. And so far this faith has been justified."[288]

3 A Belief That the Order of the World is Open to the Human Mind

Scientists assume there is an order and rationality behind the universe that science studies and at the same time they assume that the human mind is able to access and understand that rationality. According to philosopher of science Roger Trigg, "an absolute presupposition of science is the human ability to recognize what is true and reason about what could be true." This is a metaphysical presupposition because it necessarily precedes the study of the nature of the world. "Rationality and the human freedom to exercise it make scientific investigation and argument possible."[289] Without a firm conviction that "the form of things is intelligible, and therefore definable," there would be no point in embarking on the scientific quest to make sense of the world.[290] One would not scientifically seek to understand the world unless one already believed that the world could be understood. As physicist and theologian John Polkinghorne elaborates,

> We are so familiar with the fact that we can understand the world that most of the time we take it for granted. It is what makes science possible. Yet it could have been otherwise. The universe might have been a disorderly chaos rather than an orderly cosmos. Or it might have had a rationality which was inaccessible to us.... There is a congruence between our

287 Mariano Artigas, *The Mind of the Universe: Understanding Science and Religion* (London: Templeton Foundation Press, 2000), 44.

288 Paul Davies, "Taking Science on Faith," *The New York Times* (November 24th, 2007) 61–62.

289 Roger Trigg, *Beyond Matter: Why Science Needs Metaphysics* (West Conshohocken: Templeton, 2015), 71.

290 Michael Foster, "The Christian Doctrine of Creation and the Rise of Modern Natural Science," *Mind* 43 (1934): 446–68, 455.

> minds and the universe, between the rationality experienced within and the rationality observed without. This extends not only to the mathematical articulation of fundamental theory but also to all those tacit acts of judgment, exercised with intuitive skill, which are equally indispensable to the scientific endeavour.[291]

Physicist James Gates explains that in order to do science "one has to have a kind of faith that the universe is understandable." Science, says Gates, "is in fact a conversation, and you have to have faith that the universe is willing to have that conversation."[292] Every new scientific research venture assumes that the order present within the universe will lend itself to being understood by the human mind. Because this assumption that the universe will "talk back" is based on faith and cannot be given a scientific explanation, many scientists have found this relationship between our minds and the universe to be surprising and mysterious. Considering this metaphysical mystery, Einstein once reflected, "the most incomprehensible thing about the Universe is that it is comprehensible."[293] Indeed, remarks Trigg, "the intelligibility and intrinsic rationality of reality cannot be taken for granted" because "this is presupposed within science and cannot be given a scientific explanation." The presumed rationality and intelligibility of the cosmos is a "metaphysical fact, and the explanation for which, if there can be one, must come from beyond science."[294]

4 A Belief That the Order of the World is Contingent Rather Than Necessary

According to physicist and philosopher of science Mariano Artigas, "Science shows us an order that is both rational and contingent (that is, its laws and initial conditions were not necessary). It is the combination of contingency and intelligibility that prompts us to search for new and unexpected forms of rational order."[295] Trigg explains that "it was the constant temptation of ancient thinkers, such as Aristotle, to work out how the world had to be from first principles and to discount the need for a rigorous program of empirical observation

291 John Polkinghorne, *Science and Creation: The Search for Understanding* (London: SPCK, 1988), 20–1.

292 S. James Gates, "The Workings of Science," (AAAS, December 2016) http://www.scienceforseminaries.org/resource/the-workings-of-science/.

293 Albert Einstein, "Physics and Reality," in *Ideas and Opinions*, trans. Sonja Bargmann (New York: Bonanza, 1954), 292.

294 Trigg, *Beyond Matter*, 59.

295 Artigas, *Mind of the Universe*, 14–15.

and experiment."[296] The empirical focus of modern science contrasts with the mental and mathematical investigations of the ancient Greeks. "The genius of modern, empirical science, as compared with mere speculation about the nature of the world, is the realization that the physical world does not have to be as it is. It is contingent."[297] While *necessary* order could be discerned through pure introspective thought (like the truths of mathematics, geometry, or logic), *contingent* or dependent order can be discovered only by making experiments and through investigating what the world is really like. That which is contingent is knowable only by sense experience. There could have been a number of different ways that the universe was put together, but the only way to find out how it actually was put together is to examine it in its details and dynamics. In this way the early scientist Pierre Gassendi (1592–1655) derived from his faith in the contingency of the cosmos a "conviction that empirical methods are the only way to acquire knowledge about the natural world and that the matter of which all physical things are composed possesses some properties that can be known only empirically."[298] The concept of contingency "is essential to science because contingency demands an empirical method."[299] Yet, the contingency of the rational order of nature may not be investigated or established through empirical investigation. "The comprehensive presupposition upon which the whole contingent order of things reposes in order to be what it is ... cannot be established in any way from within the rational frame of the contingent order" itself.[300]

5 A Belief in Metaphysical Realism

To engage in scientific theorizing means presupposing that there is a real world of objective physical reality and that one can, at least to some extent, obtain information about that world, which exists independent of the mind. In other words, the attempt to gain knowledge about the world must first presuppose the existence of the world and that the world is not an illusion or virtual reality. "Metaphysical realism," says philosopher of science Nicholas Rescher, is not the

---

296 Trigg, *Beyond Matter*, 76.

297 Trigg, *Beyond Matter*, 76.

298 Margaret J. Osler, *Divine Will and the Mechanical Philosophy: Gassendi and Descartes on Contingency and Necessity in the Created World* (Cambridge: Cambridge University Press, 1994), 1.

299 Artigas, *Mind of the Universe*, 44.

300 Thomas F. Torrance, "The Transfinite Significance of Beauty in Science and Theology," in *L'art, la science et la métaphysique: Études offertes à André Mercier*, ed. Luz García Alonso, Evanghelos Moutsopoulos, and Gerhard Seel (Berne: Peter Lang, 1993), 393–418.

result of an inductive inference, but is rather "a regulative presupposition that makes science possible in the first place."[301] Metaphysical realism is "a precondition for empirical inquiry," and "a presupposition for the usability of observational data as sources of objective information."[302] In this way, says Rescher, "We do not learn or discover that there is a mind-independent physical reality, we presume or postulate it."[303] Trigg explains, "Science has to assume that it is investigating a world that has an independent existence. Otherwise it is a mere social construction reflecting the conditions of particular societies at a particular time."[304] The reality of the material world places crucial constraints on scientific theorizing, so true theories must match up with the structures and relationships already existing in nature. For science to make progress, reality as it concretely exists must be permitted to change one's previous abstract conceptions of that reality. This is why "scientific discoveries are often quite unexpected."[305] While scientific theories about the nature of reality can be falsified, realism itself, as a metaphysical affirmation cannot. As Trigg explains, "realism cannot be falsified, since the idea of falsification depends on notions of truth and falsity that assume that the world has an independent existence. There could otherwise be no reality to prove us wrong."[306]

6 A Belief in the Unity and Uniformity of the Physical Universe

The assumption that physical reality at some deep level is consistent, and that nature functions uniformly, is a fundamental presupposition of all scientific activity. "The idea of the general uniformity of nature," says Trigg, "underpins the conduct of science, and the alternative is to give up science. Discovering it by scientific means begs the question."[307] The "scientific method," explains philosopher of science Karl Popper, "presupposes *the immutability of natural processes*, or the 'principle of the uniformity of nature.'" For example,

301 Nicholas Rescher, *Scientific Realism: A Critical Reappraisal* (Dordrecht: Reidel, 1987), 126.

302 Nicholas Rescher, *Epistemology: An Introduction to the Theory of Knowledge* (Albany, NY: SUNY Press, 2003), 350.

303 Rescher, *Scientific Realism*, 126.

304 Roger Trigg, "Realism," in *Encyclopedia of Science and Religion*, ed. J. Wentzel van Huyssteen (New York: Macmillan Reference, 2003), 714. Biologist and theologian Alister McGrath writes, "There can be little doubt that most natural scientists espouse a range of opinions which are recognizably 'realist' in their core affirmations, reflecting a common commitment to the ontological finality of the natural order. Realism works." Alister McGrath, *A Scientific Theology, Volume. 2, Reality* (London: T & T Clark, 2006), 123.

305 Ian Barbour, *Religion in an Age of Science* (San Francisco: Harper & Row, 1990), 44.

306 Trigg, *Beyond Matter*, 100.

307 Trigg, *Beyond Matter*, 100.

physicists assume that the speed of light throughout the universe (where it has not been measured) is the same as the speed of light here on Earth (where it has been measured). This principle of uniformity, says Popper, is a "metaphysical faith in the existence of regularities in our world" that necessarily underpins the scientific method as a whole.[308] According to historian of science Reijer Hooykaas "it was not experience alone but also a belief in an order as yet undiscovered—that is, in a certain uniformity of nature—which played, and still plays an important role in science."[309] The assumption that the laws of nature are the same everywhere throughout the cosmos is what allows scientists to extrapolate from presently available knowledge to distant times (e.g., the past in geology and the past and future in cosmology) and to distant parts of the cosmos (e.g., in astronomy and cosmology). Without the postulated uniformity of the cosmos scientists could not make any inductive inferences or predictions.[310] Without this faith in nature's uniformity and unity, says Popper, any practical action within science, would be "hardly conceivable."[311]

These general presuppositions about the nature of reality—the orderliness and regularity of reality, the ontological *reality* of reality, the intelligibility and contingency of existent structures and entities, and the unity and uniformity of the physical universe—necessarily precede and underpin all scientific experimentation and reasoning. "For science to develop," says physicist and philosopher Peter Hodgson, "these beliefs must be held, at least implicitly, by society as a whole and by scientists themselves."[312] Modern science presupposes these beliefs "as the condition of its own possibility."[313] Such presuppositions (and others) are a priori "conditions that are necessary for the possibility of scientific activity as such, although they can be ignored by particular scientists."[314] As preconditions, they are absolutely required for science to take place and are not open to experimental confirmation or falsification by scientific experimentation. As Trigg explains, "empirical investigation cannot solve metaphysical issues, and if it tries to, it only goes around in circles."[315] As the necessary

308 Karl Popper, *The Logic of Scientific Discovery* (London: Unwin Hyman, 1990), 250.

309 Reijer Hooykaas, *Fact, Faith, and Fiction in the Development of Science* (Dordrecht: Kluwer, 1999), 11.

310 Ratzsch, "The Nature of Science," 49.

311 Popper, *The Logic of Scientific Discovery*, 250.

312 Hodgson, *Theology and Modern Physics*, 16. See also Hodgson, "Presuppositions and Limits of Science," 133–146.

313 Foster, "The Christian Doctrine of Creation and the Rise of Modern Natural Science," 447.

314 Mariano Artigas, "Three Levels of Interaction between Science and Philosophy," in *Intelligibility in Science*, ed C. Dilworth (Amsterdam: Rodopi, 1992), 123.

315 Trigg, *Beyond Matter*, 100.

conditions for the possibility of science these metaphysical presuppositions, explains Artigas, "continue to be present, not as a kind of philosophical ornament, but as a real part of science itself." When we study the presuppositions of science, says Artigas, "we are studying science itself in a strict sense."[316] And such presuppositions continue to significantly impact science today. McMullin says that, while "one might be tempted to think that regulative principles of a broadly metaphysical kind no longer play a role in the natural sciences ... even a moment of reflection about the current debates in elementary-particle theory, in quantum-field theory, and in cosmology ought to warn that this is far from the case."[317]

Theological Foundations of the General Metaphysical Presuppositions of Science

All the metaphysical presuppositions listed above, which continue to play a vital role within current science, require a certain degree of faith. Today, scientists take these philosophical assumptions for granted and their implicit faith in them need not necessarily be considered religious. Historically, however, each of these presuppositions developed within a specific religious context and all were supported by particular religious concepts within a particular religious culture. The specific religious context, within which early modern science developed, was the Christian faith as it was passed down from the European Middle Ages.[318] Historian of science John Hedley Brooke explains,

---

316 Artigas, *Mind of the Universe*, 25.

317 Ernan McMullin, *Newton on Matter and Activity* (Notre Dame: University of Notre Dame Press, 1978), 127.

318 In speaking of the philosophical presuppositions that emerge from the *Christian*, rather than the so-called "Judeo-Christian" faith, I do not intend to exclude Jewish or Islamic understandings of creation which often employ the same or similar concepts and philosophical presuppositions. I am merely contextualizing the discussion historically in order to avoid a lengthy digression regarding what, in fact, Christian, Jewish, and Islamic understandings of creation philosophically have in common. To go into the details with regard to the differences would involve a lengthy digression that is beyond the scope of this essay. For example, not all scholars would agree that the notion of *creatio ex nihilo* is explicitly assumed in the Hebrew thought of Genesis 1:1 or in even later Early Jewish conceptions of creation. However, *creatio ex nihilo* is unambiguously assumed in the earliest Christian understandings of creation. In a similar way, the general presuppositions about creation that emerge from Islam, Mormonism, and Process Thought have a great degree of overlap with Christian understandings. This overlap, however, is due to the historical dependence of these later metaphysical perspectives upon the Jewish and Christian concepts which preceded them. In other words, many of the general presuppositions of science that are found in Islam may be thought of as originally coming from a Jewish and/or Christian

"Prominent natural philosophers of the early modern period did not distinguish what we would call the scientific aspects of their work from what we would call theology. Their study of the natural world was conceived as a study of God's creation, disclosing something of the nature of God."[319] Within this cultural matrix, a number of specifically Christian theological understandings of the natural world and the human mind encouraged the development of the foundational presuppositions of science. In other words, "Christian theology provided several of the beliefs on which science is based."[320]

Physicist, philosopher, and theologian, Ian G. Barbour explains that a number of key metaphysical presuppositions of science are grounded in "the basic *theological affirmations* in the first chapter of Genesis." Among them are the convictions that: "the world is essentially good, orderly, coherent, and intelligible," that "the world is dependent on God" and thus contingent because "God is sovereign, free, transcendent, and characterized by purpose and will."

---

understanding. In addition to this, one should also note that scientific theory and practice arose independently in various Christian traditions outside of the Western European context. For example, within Medieval Christian Byzantium one sees a number of key scientific developments. Major theoretical achievements in physics, such as breaking with Aristotle's conception of the universe were achieved in the Byzantine era. And with regard to scientific practice, historian John Riddle explains, "Far from being static, conservative, and decadent—words used to describe Byzantine science—Byzantine medicine was sophisticated and significant, if not vibrant." John Riddle, *A History of the Middle Ages, 300–1500* (Lanham: Rowman & Littlefield, 2008), 124ff. The natural sciences also arose independently within the Aramaic-Syriac speaking Asian Christian context—such as at the University of Jundishapur (often referred to as the Arabic "house of wisdom"). The scientific work of these Syriac Christians greatly impacted the early development of the natural sciences within Islam. Describing the central role of Eastern Christians within the early Islamic study of nature, historian Philip Jenkins explains, "It is common knowledge that medieval Arab societies were far ahead of those of Europe in terms of science, philosophy and medicine, and that Europeans derived much of their scholarship from the Arab world; yet in the early centuries, this cultural achievement was usually Christian and Jewish rather than Muslim. It was Christians—Nestorian, Jacobite, Orthodox, and others—who preserved and translated the cultural inheritance of the ancient world—the science, philosophy and medicine." Philip Jenkins, *The Lost History of Christianity: The Thousand-*Year *Golden Age of the Church in the Middle East, Africa, and Asia—and How It Died* (San Francisco: HarperOne, 2008), 18; See also De Lacy O'Leary, *How Greek Science Passed to the Arabs* (London: Routledge, 2001); And see Russell M. Lawson, *Science in the Ancient World: An Encyclopedia* (Santa Barbara, CA: ABC-CLIO, 2004), xxi.

319 John Hedley Brooke, Margaret Osler and Jitse van der Meer, *Science in Theistic Contexts: Cognitive Dimensions* (Chicago: University of Chicago Press, 2001), ix.

320 Hodgson, *Theology and Modern Physics*, 17.

Barbour points out that "these are all assertions about characteristics of God and the world in every moment of time, not statements about an event in the past. They express ontological rather than temporal relationships."[321] Artigas explains how these presuppositions became deeply embedded within the intellectual milieu that gave rise to science:

> The development of empirical science as a self-sustaining enterprise required ... a kind of faith in the rationality of the world and also in the human capacity to know that world. In short, empirical science is possible only if our world possesses a strong kind of order and if we are capable of investigating it. Actually, after sharing the Christian faith for several centuries, Medieval and Renaissance Europe was built on a common ground that included, as a basic tenet, the doctrine of creation with all its implications: that the world had been created by an omnipotent and wise God and that, therefore, a natural order exists; that the natural order is contingent, because God's creation is free and thus the world cannot be a necessary product of God's action; that human beings, as creatures who participate in God's nature, can reach a knowledge of that natural order; and finally that owing to the contingent character of the world, in order to reach that knowledge we must not only think, but also perform experiments that allow us to know how our world really behaves.[322]

In the Late Medieval and Early Modern periods theological convictions became embodied within philosophical presuppositions and they worked together to form many of the key conceptual underpinnings of modern science. Rather than religion acting as a stumbling block to the rise of science, religion was, in fact, a cornerstone. As historian of science Edward Grant has shown, "in the Latin Middle Ages of Western Europe an intellectual environment was established that proved conducive to the emergence of early modern science." During this formative historical period a combination of cultural attitudes, institutions such as universities, and beliefs critically coalesced into what may be called the "the foundations of modern science."[323] Consider the theological origins for each of the metaphysical presuppositions listed above:

321 Ian G. Barbour, *When Science Meets Religion: Enemies, Strangers, or Partners?* (San Francisco: Harper Collins, 2013), 48.

322 Artigas, *Mind of the Universe*, 22.

323 Edward Grant, *The Nature of Natural Philosophy in the Late Middle Ages* (Washington, DC: Catholic University of America Press, 2010), ix.

1 The Goodness and Worth of the Physical Reality That God Created

The notion of the world's "goodness" is rooted in the foundational creation narrative of both Judaism and Christianity. In Genesis, God beholds the cosmos he created and asserts that "all that he had made" was indeed "good" (Gen. 1:31). The Hebrew word translated as "good" also means "beautiful." The created world here has an intrinsic value and the creatures therein "manifest in the most varied ways the power, wisdom, and goodness of God."[324] In the early Christian understanding, nature was seen as a type of "book" authored by God, and one could come to know God through reading and studying this book. The church father Augustine (354–430 CE) reflects, "Some people, in order to discover God, read books. But there is a great book: the very appearance of created things. Look above you! Look below you! Note it; read it. God, whom you want to discover, never wrote that book with ink; instead He set before your eyes the things that He had made. Can you ask for a louder voice than that?"[325] Continuing in this Augustinian train of thought, medieval theologian Hugh of St. Victor (1096–1141) develops specific techniques for the interpretation of the "text of nature." Hugh advocates the systematic investigation of the natural world "based on the general assumption that living things can be read as signs variously of God's power, wisdom and goodness." Discerning the power of God in the immensity of the created cosmos, Hugh likewise sees God's goodness and wisdom in the elegance and beauty of creatures.[326]

The idea that the world of nature is worth studying, as it entered into the practice of early modern science, is likewise historically rooted in the Jewish and Christian Genesis text. One particularly influential passage that deeply impacted the conceptual foundations of science was Genesis 2:19–20, in which Adam names the animals according to their own identities. Adam's naming of the different creatures had long been understood as his giving names to them in accordance with their particular natures and characteristics. Jews and Christians believed humans before the Fall had a deep knowledge of nature and that it was Adam's "encyclopedic knowledge that had made possible the

---

324 Artigas, *Mind of the Universe*, 330.

325 Quoted in Clarence Glacken, *Traces on the Rhodian Shore: Nature and Culture in Western Thought from Ancient Times to the End of the Eighteenth Century* (Berkeley: University of California Press, 1967), 203–204.

326 Peter Harrison, *The Bible, Protestantism, and the Rise of Natural Science* (Cambridge: Cambridge University Press, 1998), 57; Hugh of St Victor, *Didascalicon*, 6.5; *De tribus diebus* 1.

naming" of the various animals.[327] In the 1600s, when Francis Bacon inaugurated the modern scientific endeavor, he drew upon this understanding of Adam's knowledge of the natural world. Bacon envisioned the natural sciences as a way of "restoring, or at least repairing, the losses to knowledge that had resulted from the Fall."[328] Historian of science Peter Harrison explains, "Francis Bacon's project to reform philosophy was motivated by an attempt to determine whether the human mind 'might by any means be restored to its perfect and original condition, or if that may not be, yet reduced to a better condition than that in which it now is.'"[329] As the disobedience of the first humans caused the human mind to fall into error and lose knowledge, the scientific method was, for Bacon and other early modern practitioners of science, a technique that could work to heal the cognitive damage wrought by human sin. During the scientific revolution, says Harrison, "the methodological strictures of particular programs of natural philosophy—experimental method being perhaps the best example—were understood as applying necessary external constraints to fallen minds which, left to their own devices, would simply fail to accumulate any useful knowledge of the natural world."[330]

2 That God Created an Orderly and Rational Cosmos

The orderliness and rationality of the natural world were similarly assumed by early modern scientists on the basis of the Christian doctrine of creation that was part of their cultural matrix.[331] "The very idea of rationality has certain theological origins, and science as we know it arose in the context of a belief in the rational structure of reality mirroring the higher wisdom of a Creator God."[332] The concept of God's creation of all material reality out of nothing (Latin: *creatio ex nihilo*) "allowed the scientist to approach nature with the expectation that the divine rationality would be reflected in its structures and workings."[333] According to Hooykaas, "The faith in order, law, simplicity,

327 Peter Harrison, *The Fall of Man and the Foundations of Science* (Cambridge: Cambridge University Press, 2007), 26.

328 Harrison, *The Fall of Man and the Foundations of Science*, 4.

329 Harrison, *The Fall of Man and the Foundations of Science*, 1.

330 Harrison, *The Fall of Man and the Foundations of Science*, 15.

331 Though anachronistic, the word "scientists" is used here for clarity's sake.

332 Carl Reinhold Bråkenhielm, "Theology and the Origins of Customized Science" in *The Customization of Science: The Impact of Religious and Political Worldviews on Contemporary Science*, ed. Steve Fuller, Mikael Stenmark and Ulf Zackariasson (London: Palgrave Macmillan, 2014), 121.

333 Alister McGrath, *A Scientific Theology, Volume 1, Nature* (Edinburgh: T & T Clark, 2003), 140.

harmony, beauty has often been connected with the faith that there is logos, reason, mind at work in the universe." The idea that the universe is deeply rational emerges from a "belief in a Mind to which the human mind has, however remotely, some resemblance, so that it is able to recognize these attributes in a creation which is the work of that Mind."[334] Jews and Christians believed that the ways of nature, as the product of the Divine Mind, were reflections of reason and that "even those aspects of nature that threatened human safety were not lawless in themselves. They served God's purposes and had laws of their own, even if unknown to humans (Job 28:25–27)."[335]

3 That God Created the Human Mind to Comprehend God's Cosmos

Since God's creative activity in the cosmos reflects the rationality of the Divine Mind, Christians believe that the inner workings of the cosmos "are open to human comprehension, at least in principle."[336] As historian of science Christopher Kaiser explains, "The creation of all things by God, the consequent order and rationality of the cosmos, and the ability of human reason to comprehend this order all stem from the Judeo-Christian belief in creation, dating back at least to the second century BCE."[337] In this way, says theologian Alister McGrath, "human rationality thus bears a created, contingent relationship

334 Hooykaas, *Fact, Faith, and Fiction*, 12. Cambridge historian of science Joseph Needham, the greatest Western interpreter of the history of science in China, has argued that this Judeo-Christian metaphysical faith in the rationality of the Creator was one of the key reasons that theoretical science developed and flourished in a European context in a way that it never did in a Chinese context—even though technology in China was well-developed. Needham explains that in China, "the highest spiritual being known and worshipped was not a Creator in the sense of the Hebrews and the Greeks. It was not that there was no order in Nature for the Chinese, but rather that it was not an order ordained by a rational personal being, and hence there was no guarantee that other rational personal beings would be able to spell out in their own earthly languages the pre-existing divine code of laws which he had previously formulated. There was no confidence that the code of Nature's laws could be unveiled and read, because there was no assurance that a divine being, even more rational than ourselves, had ever formulated such a code capable of being read." Joseph Needham, *Human Law and the Laws of Nature in China and the West* (L.T. Hobhouse Memorial Trust Lecture, Cambridge University Press, 1951), 41–2.

335 Christopher B. Kaiser, "Early Christian Belief in Creation and the Beliefs Sustaining the Modern Scientific Endeavor," in *The Blackwell Companion to Science and Christianity*, ed. J.B. Stump and Alan G. Padgett (Malden, MA: Blackwell, 2012), 6.

336 Kaiser, "Early Christian Belief in Creation, 6.

337 Kaiser, "Early Christian Belief in Creation, 10.

to–but is not identical with–divine rationality."[338] Affirming that the natural world could be comprehended, "early Christian scientists sought intelligible order in nature, regarding it as an indication of God's rational plan for the universe."[339]

4 The Created Contingency of the Cosmic Order

According to the Christian theological context within which the natural sciences developed, "God is the creative ground and reason for the contingent but rational unitary order of the universe."[340] The "Christian doctrine of creation" affirms that "the universe is both inherently intelligible and inherently contingent, its intelligibility reflecting its contingent origins in the rationality of God."[341] The belief that the order of the world is contingent rather than necessary is ultimately grounded in the Christian conception of the freedom of God.[342] Inherent in the Christian doctrine of *creatio ex nihilo*, which provided the conceptual matrix for early modern science, is the belief that God was free to choose how to create the universe. "Biblical thought held that the world's order is contingent rather than necessary. If God created both form and matter, the world did not have to be as it is, and one has to observe it to discover the details of its order."[343] God "was not in any way constrained either to create or not to create it in the way that He did. It is therefore not a necessary universe in the sense that it had to be created or could not have been created otherwise."[344] Given this understanding of nature, one can never say a priori (independently of observation) how God *must have* acted, and thus one can never say a priori how God's creation must behave. To obtain true knowledge about God's creation one must proceed in an *a posteriori* manner—by studying

338 Alister E. McGrath, *The Open Secret: A New Vision for Natural Theology* (Malden, MA and Oxford: Blackwell-Wiley, 2008), 192.

339 Paul Davies, "The intelligibility of nature," in *Scientific Perspectives on Divine Action: Quantum Cosmology and the Laws of Nature*, ed. Robert J. Russell, Nancey Murphy, and C.J. Isham (Berkeley, CA: CTNS and Vatican Observatory Publications, 1999), 149–164.

340 Artigas, *Mind of the Universe*, 15.

341 McGrath, *Open Secret*, 237.

342 Kaiser, "Early Christian Belief in Creation," 5. McGrath writes, "The creator is to be regarded as free of limitations imposed by the 'inertia of a prior reality.'" McGrath, *A Scientific Theology*, 1,195.

343 Barbour, *When Science Meets Religion*, 48.

344 Hodgson, *Theology and Modern Physics*, 26. See also Osler, *Divine Will and the Mechanical Philosophy*.

the material creation and by conducting experiments.[345] Thus early scientists such as "Gassendi described a world utterly contingent on divine will. This contingency expressed itself in his conviction that empirical methods are the only way to acquire knowledge about the natural world and that the matter of which all physical things are composed possesses some properties that can be known only empirically."[346] More recently, the essential affirmation of the contingency of the cosmic order "can be seen as lying behind both James Clerk Maxwell's insistence that there exists an inner relation between the laws of the mind and the laws of nature, and Albert Einstein's belief in a 'pre-established harmony' between the intelligibility of the independent world and the perceiving subject."[347]

5 The Independent Reality of the Created Cosmos

Related to the contingency of the order in the physical world is the notion of metaphysical realism.

Metaphysical realism is grounded in the Jewish affirmation that God created the natural world to possess a significant degree of relative autonomy or independence.[348] Inheriting the metaphysical framework of its Jewish forebearers, the Christian theological tradition clearly presupposed and applied a philosophical faith in the existence of the external world, which has a structure that is independent from the human mind.[349] In the Jewish and Christian understandings, the material creation exists independently of the observer because God the creator exists and bestowed existence on both the human observer and the created objects being observed. The reality of both the external world and the human observer are affirmed because they are the creation of the same God.[350] The material world is understood as having its own reality owing to the fact that creation is independent or distinct from the Creator.[351]

"On the Christian conception ... nature is made by God, but *is not God.* There is an abrupt break between nature and God. Divine worship is to be paid to God alone, who is *wholly other than nature.* Nature is not divine."[352] Barbour

345 Michael Foster, "Greek and Christian Ideas of Nature," *The Free University Quarterly* 6 (1959), 125; McGrath, *Scientific Theology*, 2.139.

346 Osler, *Divine Will and the Mechanical Philosophy*, 1.

347 McGrath, *The Open Secret*, 237.

348 Christopher Kaiser, "Early Christian Belief in Creation," 7.

349 McGrath, *Scientific Theology*, 2.199.

350 McGrath, *Scientific Theology*, 2.172, 228.

351 "The reality of God and the derived and contingent reality of the creation can thus be seen as distinct." McGrath, *Scientific Theology*, 2.228.

352 Foster, "Greek and Christian Ideas of Nature," 123–24; See also Artigas, *The Mind of the Universe*, 22.

refers to this presupposition about the independent reality of the cosmos as a belief in the dedivinization or desacralization of nature.[353] As the independent creation of God, "the Christian cosmos is not inhabited by deities. Yet, as a divine creation it does bear deep theological significance."[354] Within the religions of the ancient word, Christians and Jews were unique in their denial of the divinity of the celestial bodies and "this skepticism was motivated by a theological worldview."[355] Because of this disbelief in the divinity of the sun, moon, earth, planets, and stars, Christians were often labeled as atheists and they were lumped together with the Epicureans who did not believe in any gods at all. The Neoplatonist philosopher Celsus and other educated pagans ridiculed Jews and Christians for their impiety in this matter. The atheist reputation of Christians continued into late antiquity with the Aristotelian philosopher Simplicius, "being horrified at the blasphemy" of the Christian philosopher John Philoponus (ca. 490–570), "who denied divinity to heavenly bodies."[356] Following the thought of Basil of Caesarea, Philoponus' theology led him to believe "that the motion of the heavens was to be explained by a 'motive force' imparted by God at the moment of creation."[357] Philoponus's view of the independence and reality of the creation "supposed a unified theory of dynamics" where all natural motion was imparted upon creation by God. Philoponus's conception of impetus subsequently influenced Galileo and all those early scientists who would follow in his footsteps.

6 The Unity of Creation as Grounded in the Unity of God

The affirmation of the unity and uniformity of the physical universe was likewise a core belief emerging from a Judeo-Christian understanding of the unity of creation as the product of a single Creator. While many ancient schools of thought "drew a sharp line between the starry heavens and the terrestrial realm," the Christian tradition insisted on "a single physics for both heaven and earth."[358] This conception of the cosmos had become well established in the early church and was passed down to later Islamic and medieval Christian thinkers. It was then handed on from the leading natural philosophers of the Middle Ages to the practitioners of early modern science. When early modern scientists, such as Isaac Newton, argued for the universality of the laws of

353 Barbour, *When Science Meets Religion*, 48.

354 Harrison, *Territories*, 53.

355 Harrison, *Territories*, 53.

356 Harrison, *Territories*, 53.

357 Harrison, *Territories*, 53.

358 Christopher Kaiser, "The Creationist Tradition in the History of Science," *Perspectives on Science and Christian Faith* 45 (June 1993), 80–89.

nature they justified this principle in theistic terms. Newton says, "If there be a universal life and all space be the sensorium of a thinking being [(God)] who by immediate presence perceives all things in it, [then] the laws of motion arising from life or will may be of universal extent."[359] In the nineteenth century, the "quest for a unification of electricity, magnetism, and optics, culminating in the work of James Clerk Maxwell, was still inspired by this theological ideal."[360] Theological presuppositions about the unity of creation also clearly motivated Michael Faraday in his scientific quest to discover the fundamental principles underlying electromagnetism and electrochemistry. As historian of science Colin Russell says, "No doubt Faraday's belief in the unity of the forces of matter was reinforced by his faith in a Creator who made the whole universe work together in harmony."[361] The influence of this theological affirmation also played a vital role in the development of cosmological theory in the 20th century. According to Brooke, "the inculcation of a Jewish monotheism early in life had a lasting effect in the way Einstein was driven, as many physicists still are, to seek a theory that would unify the fundamental physical forces."[362]

Is Ontological Naturalism a General Presupposition of Science?

In our current culture many believe that atheism has a privileged position within the conceptual framework of the natural sciences. "It is certainly the case," says philosopher of science Michael Ruse, "that a lot of people, perhaps Richard Dawkins most famously, think that science implies atheism."[363] Some have gone beyond this to argue that science necessarily demands an atheistic worldview and can only properly be done within an atheistic intellectual

359 Ratzsch, "The Nature of Science," 65. A *sensorium* is the sum of an organism's perception, the "seat of sensation" where it experiences and interprets the environment it lives within.

360 Kaiser, "The Creationist Tradition in the History of Science," 80. McGrath observes, "Since the uniformity of nature is an unjustified (indeed, circular) assumption within any non-theistic world-view, it could be argued that there is no firm basis upon which to engage in scientific activities, other than the belief that the regularities observed locally prove universal." McGrath, *Scientific Theology*, 2.153.

361 Colin Russell, *Michael Faraday: Physics and Faith* (New York: Oxford University Press, 2000), 105.

362 John H. Brooke, "Preface," in *Science in Theistic Contexts: Cognitive Dimensions, Osiris 16*, ed. John H. Brooke, M.J. Osler, J.M. Van der Meer (Chicago: University of Chicago Press, 2001), viii.

363 Michael Ruse, "Atheism and Science," in *The Customization of Science: The Impact of Religious and Political Worldviews on Contemporary Science*, ed. Steve Fuller, Mikael Stenmark and Ulf Zackariasson (London: Palgrave Macmillan, 2014), 74.

framework. Thus historian Tor Egil Førland argues that "The scientific worldview has no room for God or other supernatural entities," and that scientists' "explanations of how the world works have no room for supernatural beings."[364] In this way Førland and others have argued that science must be founded upon ontological naturalism. According to philosopher of science Ronald Giere ontological naturalism "implies the rejection of supernaturalism ... the rejection of any deity such as the Judeo-Christian God, which stands outside nature as creator or actor. Positively, naturalists hold that reality, including human life and society, is exhausted by what exists in the causal order of nature."[365]

Does science necessarily entail ontological naturalism? Philosopher René van Woudenberg points out that ontological naturalism "is a very weighty substantial metaphysical assumption" that comes at a high cost.[366] By excluding all entities and phenomena "outside the causal order of nature," one would exclude not only a Creator God, but also "logical and moral principles, numbers, sets and other abstract objects."[367] None of these would be considered as existing. But, asks van Woudenberg, "what reason could one have to adopt this very weighty metaphysical claim?" If this claim really is an *assumption* of science, "reasons for accepting it cannot emerge from scientific investigation. That would be moving in a circle."[368] First one would have to assume "that there is no God, no creator, nor anything outside the causal order of nature; and upon noticing that science offers no arguments for the existence of God or creator or any of the other items, one concludes that there is no God, no creator nor any of the other items." The circular logic here, says van Woudenberg, "is blatantly obvious." Thus, any genuine basis for ontological naturalism must be found outside of science. But if this is the case, then the argument of the ontological naturalist is self-defeating because "the typical naturalist won't accept that outside science anything of worth is to be found!" Philosophically speaking, "acceptance of epistemological naturalism is tantamount to intellectual suicide and hence cannot be considered as an assumption that underlies science."[369]

364 Tor Egil Førland, "Acts of God? Miracles and Scientific Explanation," *History and Theory* 47 (December 2008), 483–494, 484, 492.

365 Ronald N. Giere, "Naturalism," in *A Companion to the Philosophy of Science*, ed. W.H. Newton-Smith (Oxford: Blackwell, 2000), 308.

366 René van Woudenberg, "Christianity and Science," in *The Customization of Science: The Impact of Religious and Political Worldviews on Contemporary Science*, ed. Steve Fuller, Mikael Stenmark and Ulf Zackariasson (London: Palgrave Macmillan, 2014), 63.

367 Van Woudenberg, "Christianity and Science," 63.

368 Van Woudenberg, "Christianity and Science," 63.

369 Van Woudenberg, "Christianity and Science," 64.

The claim that ontological naturalism is a general metaphysical presupposition of science is problematic for historical reasons as well. Most scientists throughout time have not assumed ontological naturalism and many have explicitly rejected such naturalism. "From a historical point of view," says van Woudenberg, "it needs to be stressed that many of the best scientists would be hostile to the very suggestion" that science is based on the presupposition of ontological naturalism. A few prominent scientists who have explicitly rejected ontological naturalism include Copernicus, Kepler, Galileo, Steno, Newton, Boyle, Darwin, Mendel, Sedgwick, Plank, Dobzhansky, Faraday, Maxwell, Lemaitre, Bohr, and Einstein. Since these persons have significantly shaped the course of science for over 400 years, "it seems unlikely that *they* were mistaken about what science presupposes!"[370]

A third problem is that a science that assumes ontological naturalism places an non-natural limit on the scope of scientific discovery. "Science, by its very nature, is an open-ended affair, it goes, and ought to go, wherever the evidence leads." Thus, asks van Woudenberg, "How can one assume in advance that science will never lead in the direction of a God, or a creator, or of items that are causally inert?" At some point in the history of science there might be very good theoretical reasons for the existence of such things. Given these problems and others, Ruse concludes that "the arguments that are given for suggesting that science necessitates atheism are not convincing."[371] Science, says Ruse, "can never, even in principle, be metaphysically atheistic," and "the direction of modern science is such that ultimately no argument can be mounted on the grounds of science alone that there is no God."[372] In the end there is no compelling reason to embrace the idea that ontological naturalism is a metaphysical presupposition of science. As van Woudenberg says, "there is no reason to think that naturalism either is or needs to be considered as an assumption that underlies science. And, hence, naturalism should not be thought of as an assumption that uniquely characterizes science."[373]

Particular and Contextual Metaphysical Presuppositions of Science

In addition to the *general* philosophical presuppositions that are provided—both historically and presently—there are also *particular presuppositions* or metaphysical assumptions that correspond to particular *metaphysical frameworks* and *research paradigms* within which scientific theorizing may take

370 Van Woudenberg, "Christianity and Science," 64.

371 Michael Ruse, "Atheism and Science," 75.

372 Ruse, "Atheism and Science," 75–76.

373 Van Woudenberg, "Christianity and Science," 62.

place.[374] These particular metaphysical assumptions and paradigms play a more circumscribed role within scientific theorizing than the general presuppositions and are often related to certain stages in the historical development of a given scientific theory.

For example, the commandment to humans in Genesis to exercise dominion over nature played an important role in the rise of Early Modern science that it no longer plays today. In the Middle Ages many believed "that Adam's original dominion over the creatures in the Garden of Eden consisted in a mental mastery of what it was that they represented."[375] One consequence of the Fall was that this original knowledge (or *science*) was lost, and the powers of the mind—"sense, imagination, reason, understanding, intelligence, and moral discernment—were distorted by sin." The recapturing of this lost mastery and knowledge of nature "could be achieved only if the powers that had originally made it possible were 'cleansed by righteousness, trained by learning, and perfected by wisdom.'"[376] The scientific enterprise that began in the late medieval period was aimed at gaining knowledge of the natural world in order to restore to the human mind some of its original powers and perfections. In this way, says Harrison, the biblical injunction to exercise dominion, "underpinned the modern scientific project, providing an important source of motivation for the investigation of nature and giving religious legitimacy to a project that ... was more vulnerable in its early stages than we have sometimes assumed."[377]

As another example, consider the period when the geological sciences were first developing in 17th and 18th century Europe.[378] At this time one important particular presupposition that was a matter of significant debate between geologists was related to the question of whether the planet Earth had a beginning and changed progressively through time, or alternatively, whether Earth was *infinitely old* (without beginning or end) and characterized by non-progressive cyclical geological processes.[379] The first view is known as Historical Geology and the second view—an idea that had been promoted by Aristotle and other

374 Scientific research paradigms are "standard examples of scientific work that embody a set of conceptual and methodological assumptions." See Barbour, *Religion in an Age of Science*, 51.

375 Peter Harrison, *Territories*, 65.

376 Harrison, *Territories*, 66.

377 Harrison, *Territories*, 137.

378 For a detailed discussion see Joshua M. Moritz, *Science and Religion Beyond Warfare and Toward Understanding* (Winona, MN: Anselm Academic, 2016), 44–48.

379 Alan H. Cutler, "Nicolaus Steno and the Problem of Deep Time," in *The Revolution in Geology from the Renaissance to the Enlightenment*, ed. Gary D. Rosenberg (Boulder, CO: Geological Society of America, 2009), 143–48, 143.

ancient thinkers—is Aristotelian Eternalism (or Geologic Eternalism). Before the relevant empirical evidence was available to decide between the two alternative theories of time, history, and progressive change as it related to Earth's processes, discussions among early geologists about the timescale of the world was deeply colored by a "clash of theologies." As historian of geology Martin Rudwick explains, this "was not a case of 'Religion versus Science,' but of one religious view of the world against another."[380] At that point in time, there was not conclusive empirical evidence to demonstrate whether the planet Earth had a physical beginning or whether it was, in fact, eternal. The working assumption that Earth was a historical entity thus served as a particular presupposition that, as evidence was accumulated in its favor, would—in time—no longer be a matter of philosophical or theological faith.

A third example of a particular presupposition is found in Charles Darwin's context of discovery. At the core of Darwin's scientific quest to establish the common ancestry of all life was a *theologically* and *morally* inspired conviction in the ancestral unity of humanity.[381] This conviction motivated and drove Darwin's research agenda. According to Darwin's biographers, Adrian Desmond and James Moore, "Rather than seeing 'the facts' force evolution on Darwin, "we find a *moral* passion firing his evolutionary work. He was quite unlike the modern 'disinterested' scientist who is supposed (supposed, mark you) to derive theories from 'the facts' and only then allow the moral consequences to be drawn."[382] The notion of human unity—along with the corresponding rejection of slavery—was a key element of Darwin's family heritage. "Adamic unity and the brotherhood of man were axiomatic in the anti-slavery tracts that he and his family devoured and distributed. It implied a single origin for black and white, a shared ancestry."[383] When Darwin began his evolutionary quest in search of human origins, his "starting point was the abolitionist belief in blood kinship, a 'common descent'" for all human beings. And this deep conviction and faith in the unity of the human race "was the *unique* feature of Darwin's peculiar brand of evolution."[384] As evidence for the common ancestry of

380 Martin Rudwick, *Bursting the Limits of Time: The Reconstruction of Geohistory in the Age of Revolution* (Chicago: University of Chicago Press, 2005), 118.

381 For a detailed discussion see Joshua M. Moritz, *Science and Religion: Beyond Warfare and Toward Understanding* (Winona, MN: Anselm Academic, 2016), 48–51.

382 Adrian Desmond and James Moore, *Darwin's Sacred Cause: How a Hatred of Slavery Shaped Darwin's Views on Human Evolution* (Boston: Houghton Mifflin Harcourt, 2009), xviii.

383 Desmond and Moore, *Darwin's Sacred Cause*, 54, xvii.

384 Desmond and Moore, *Darwin's Sacred Cause*, xvii.

humanity and all of life was increasingly found, the particular presupposition of Adamic Unity became less important as a guiding principle.

While the above examples are all particular presuppositions emerging from a theistic context, there are in the history of science "metaphysical motivations of an antireligious nature" that have guided research programs as well.[385] For instance in the 20th century the atheistic presuppositions of Cambridge Cosmologist Fred Hoyle motivated him to develop Steady State Theory—according to which the universe has always existed and will continue to exist in a state similar to that which we perceive now. Even as Hoyle's atheism was a key motivator in the development of this alternative theory, it also motivated his lifelong refusal to accept the big bang theory of Georges Lemaître. "Hoyle disliked the notion of an initial cause beyond the realms of science, which is what seems to be implied by the Big Bang," explains historian and physicist Rodney Holder, "and he certainly associated the steady-state theory with atheism."[386] In other words, says historian of science Edward Davis, "Precisely because he did not believe in God, and because he did not like the implications of Lemaître's theory, Hoyle ridiculed" it and with "two other English astronomers proposed the 'steady-state' theory, to provide an alternative explanation without the implication that the universe had a beginning."[387]

Another example of particular metaphysical presuppositions of an atheist kind is found in the work of the mathematician Jean Paul Van Bendegem. Van Bendegem rejects the idea of the infinite and endorses mathematical finitism.[388] Van Bendegem's "motivation for taking this position is that acceptance of the infinite would imply or suggest acceptance of the existence of God."[389]

In a similar way Stanford philosopher of science Nancy Cartwright has been led to reject the notion of laws of nature on account of this concept's implicit theological ramifications. Cartwright says, "I think that in the concept of law there is a little too much of God.... In the end the concept of a law does not

385 Van Woudenberg, "Christianity and Science," 66.

386 Rodney Holder, "Georges Lemaître and Fred Hoyle: Contrasting Characters in Science and Religion," in *Georges Lemaître: Life, Science and Legacy*, ed. Rodney Holder and Simon Mitton (New York: Springer, 2013), 46.

387 Edward Davis, "Is the Big Bang Theory Irreligious?," *Harrisburg Patriot-News* (25 July 2002), A-13. See also Helge Kragh, *Cosmology and Controversy: The Historical Development of Two Theories of the Universe* (Princeton: Princeton University Press, 1999), 192ff.

388 Jean Paul Van Bendegem, "Why the Largest Number Imaginable is Still a Finite Number," *Logique et Analyse*, 42 (2002), 107–26.

389 Van Woudenberg, "Christianity and Science," 66.

make sense without the supposition of a law-giver."[390] One of Cartwright's predominant reasons for opposing the concept of laws of nature is "the difficulty of incorporating the laws of nature in an atheist world-view." This difficulty, explains Cartwright "is why I have been combating laws of nature all these years."[391] Since, according to Cartwright, "we cannot understand natural law without resorting to God," she argues that scientists must endeavor "to take into account modern science, without bringing the idea of law into play."[392]

A current example of particular metaphysical presuppositions that play an active role in guiding a scientific research program is found in Chinese paleoanthropology as it is influenced by ethnic nationalism. According to paleoanthropologist Jennie Jin, "Ethnic nationalism" is "a potent force in shaping modern day Chinese paleoanthropology."[393] The understanding of race in Chinese racial nationalism "requires rejection of a key theory of international paleoanthropology—the biological unity of humanity, which stems from its common origin in Africa."[394] The majority of Chinese paleoanthropologists reject "the international mainstream palaeoanthropological theory denoting Africa to be the origin of both the hominid and all modern humans"[395] and instead, they advocate a "scientific Sinocentrism, or paleontological nationalism, that bolsters China's deep history back to the time of the Peking Man."[396] The guiding assumption of ethnic nationalism has motivated "many Chinese scientists and the general public" to "hold the interpretation that modern Chinese are descendants of Peking Man."[397] Yet, metaphysical presuppositions are especially difficult to discern when they are your *own* key assumptions. Thus Chinese researchers reject that their interpretation of the data is being guided by ethnic nationalism. As Wu Xinzhi, a paleontologist at the Chinese Academy

390 Nancy Cartwright, "Is Natural Science "Natural" Enough? A Reply to Philip Allport," *Synthese* 94 (1993): 291–301, at 299.

391 Nancy Cartwright quoted in Lydia Jaeger, "Nancy Cartwright's Rejection of the Laws of Nature and the Divine Lawgiver," *Science and Christian Belief* 22:1 (2010), 86.

392 Jaeger, "Nancy Cartwright's Rejection of the Laws of Nature," 86.

393 Jennie Jin , "The People's Peking Man: Popular Science and Human Identity in Twentieth-Century China," Book Review, *PaleoAnthropology* (2010): 72–77, 73.

394 Barry Sautman, "Peking Man and the Politics of Paleoanthropological Nationalism in China," *The Journal of Asian Studies* 60:1 (Feb., 2001), 95–124, 96.

395 Hsiao-pei Yen, "From palaeoanthropology in China to Chinese palaeoanthropology: Science, imperialism and nationalism in North China, 1920–1939," *History of Science* 53:1: (2012) 21–56, 22.

396 Hsiao-pei Yen, "Evolutionary Asiacentrism, Peking Man, and the Origins of Sinocentric Ethno-Nationalism," *Journal of the History of Biology* 47 (2014): 585–625, 591.

397 Hsiao-pei Yen, "Evolutionary Asiacentrism," 590.

of Sciences' Institute of Vertebrate Paleontology and Paleoanthropology says, "This has nothing to do with nationalism ... It's all about the evidence—the transitional fossils and archaeological artefacts ... Everything points to continuous evolution in China from *H. erectus* to modern human."[398]

Whether scientists are aware of them or not, particular presuppositions never disappear from science. All data collected via the scientific method and the interpretation of this data is dependent upon the particular research paradigm within which that data is considered. Such data is never free of the theoretical assumptions related to a given paradigm and the choice of one research paradigm over another is not dictated by scientific research in itself.[399] Unless a scientist is so radically aware of his or her own metaphysical presuppositions that he or she can systematically strive to replace them with other presuppositions, scientific theories will inevitably be formulated and interpreted in such a way as to achieve consistency with dominant metaphysical presuppositions.

Does Science have Limits?

Are the natural sciences *omni-competent*? Can science answer all possible questions? Or are there limits to what science can understand and explain? The enormous success of the scientific enterprise since the 1700s has instilled in the minds of many an unyielding confidence that empirical methods will eventually address any and all questions worth asking. The logical empiricist philosopher Rudolph Carnap maintained, "When we say that scientific knowledge is unlimited, we mean that there is no question whose answer is in principle unattainable by science."[400] Carnap affirms that in time, science can account for everything that is and everything that will be in the cosmos. Oxford Chemist Peter Atkins shares Carnap's optimism as he declares that "science is slowly equipping itself to deal with aesthetic and religious experiences, and will be able to account for the perception of oneself as a distinct but responding entity. It will do so, I do not doubt, by showing that these characteristically human capacities, which we lump together for convenience of discourse as 'human spirit' or 'soul', are no more than psychological states of the brain."[401]

398 J. Qiu, "How China Is Rewriting the Book on Human Origins," *Nature* 12:535 (July 2016): 218–220, 218.

399 See Barbour, *Religion in an Age of Science*, 52.

400 Rudolph Carnap, *The Logical Structure of the World* (London: Routledge and Kegan Paul, 1967), 290.

401 Peter Atkins, "The Limitless Power of Science," in *Nature's Imagination: The Frontiers of Scientific Vision*, ed. John Cornwell (Oxford: Oxford University Press, 1995), 122–132, 129.

The assertions of Carnap and Atkins well express a view called *scientism*—the belief "that any question that can be answered at all can best be answered by science."[402] Philosopher Mikael Stenmark describes scientism as the belief that "there are no real limits to the competence of science, no limits to what can be achieved in the name of science. There is nothing outside the domain of science, nor is there any area of human life to which science cannot successfully be applied. A scientific account of anything and everything constitutes the full story of the universe and its inhabitants."[403] Scientism is an overarching perspective that is quite common among atheist and agnostic scientists, philosophers, and promoters of science. For example, Richard Dawkins, evolutionary biologist and public advocate of both science and atheism, believes that the natural sciences—and in particular modern biology—will deliver answers to all of life's big questions. Dawkins holds that progress in science will, in time, lead to a place where people have "no longer ... to resort to superstition when faced with the deep problems: Is there a meaning to life? What are we for? What is man?"[404]

Another atheist biologist, the late Stephen Jay Gould, affirms that even now science can provide a clear answer to the perennial question "Why are we here?" Gould responds that humans "are here because one odd group of fishes had a peculiar fin anatomy that could transform into legs for terrestrial creatures; because the earth never froze entirely during an ice age; because a small and tenuous species, arising in Africa a quarter of a million years ago, has managed, so far, to survive by hook and by crook." Although, says Gould, "we may yearn for a higher answer ... none exists."[405] This is because, advocates of scientism argue, any knowledge that is not ultimately reducible to empirical testing is essentially meaningless. All knowledge that is true knowledge, biologist E.O. Wilson confidently contends, and every enterprise of culture—even the humanities and creative arts—will eventually be "subsumed by the natural sciences."[406] As science expands to encompass all domains of knowledge, affirms atheist philosopher Daniel Dennett, the ultimate reasons lying behind

402 John *Dupré, Human Nature and the Limits of Science* (Oxford: Oxford University Press, 2001), 2.

403 Mikael Stenmark, "Science and the Limits of Knowledge," in *Clashes of Knowledge: Orthodoxies and Heterodoxies in Science and Religion*, ed. Peter Meusburger, Michael Welker, and Edgar Wunder (New York: Springer, 2008), 111.

404 Richard Dawkins, *The Selfish Gene* (Oxford: Oxford University Press, 1989), 1.

405 Stephen Jay Gould, quoted in Michael Ruse, *Science and Spirituality: Making Room for Faith in the Age of Science* (Cambridge: Cambridge University Press, 2010), 6.

406 Edward Wilson, *Consilience: The Unity of Knowledge* (New York: Knopf, 1999), 12.

"everything we value—from sugar and sex and money to music and love and religion" will be revealed as "evolutionary reasons, free-floating rationales that have been endorsed by natural selection."[407] There are no things science won't eventually know. There are no limits to the scientific enterprise, and nothing is, in principle, beyond the reach of scientific understanding.

In evaluating the perspective of scientism, one might ask if such unswerving confidence in the power of science is, itself, strictly scientific. E.O. Wilson admits that it is not, and that the belief that all knowledge will one day be unified under the natural sciences is "a metaphysical world view" with "its best support" being "no more than an extrapolation of the consistent past success of the natural sciences."[408] In other words, scientism is based on a non-empirical *faith* in the future success of science. One might also ask if proponents of scientism are justified in claiming that *all* legitimate questions about the meaning and purpose of life will in time be answered by the natural sciences. The empirically-minded rationalist philosopher Peter Medawar believes not: "That there is indeed a limit upon science is made very likely by the existence of questions that science cannot answer, and that no conceivable advance of science would empower it to answer." Questions such as "What are we all here for?" or "What is the point of living?" Medawar believes are forever beyond the purview of science.[409] Nor does it seem reasonable—or scientific—to declare that all answers that are not scientific are consequently meaningless. Indeed, the claim that all non-scientific explanations are meaningless is a logically self-defeating claim. As philosopher of science Mariano Artigas explains, "The claim that there is no valuable knowledge *outside* science certainly cannot be supported from *within* science. This is an extremely simple philosophical error, akin to a child claiming that because all the people he knows are in his house, that there cannot be any people outside his house."[410] According to Artigas, the epistemology (or theory of knowledge) that scientism offers is "not the consequence of any scientific investigation but rather reaches outside itself into the very realm that it claims does not exist."[411] Because the ultimate foundation

407 Daniel Dennett, *Breaking the Spell: Religion as a Natural Phenomenon* (New York: Viking, 2006), 93.

408 Wilson, *Consilience*, 9.

409 Alister E. McGrath, "The Ideological Uses of Evolutionary Biology in Recent Atheist Apologetics," in *Biology and Ideology from Descartes to Dawkins*, ed. Denis R. Alexander and Ronald L. Numbers (Chicago: University of Chicago Press, 2010), 340.

410 Mariano Artigas and Karl Giberson, *Oracles of Science: Celebrity Scientists versus God and Religion* (Oxford: Oxford University Press, 2007), 40.

411 Artigas and Giberson, *Oracles of Science*, 40.

for scientism is metaphysical rather than strictly empirical, "the intellectually unjustified misstep that provides the foundation for scientism ... derives not from science, but from the transgression of science's own self-imposed limits."[412] Consequently, says McGrath, "It is clear that there are many questions that, by their very nature, have to be recognized as lying beyond the legitimate scope of the scientific method."[413]

Another problem with scientism is that it implicitly rejects scientific knowledge, pointing to areas where empirical research has discovered limits to what can be known through observation and experiment. Consider, for example, Big Bang cosmology—the leading scientific theory that describes the origin, structure, and development of the universe. According to the Standard Big Bang model, which is derived both from Einstein's theory of general relativity and observational data, the universe began 13.7 billion years ago in a *singularity*—an infinitely small point in which matter was infinitely compressed. Everything that physically exists, including matter, energy, space, and time, came into existence at the big bang singularity, and thus it makes no sense to speak of physical reality or even a "time before" this point. The existence of an initial singularity of this sort represents a fundamental limit to observational powers of science.[414] Within a fraction of a second after the initial singularity began to expand, the unimaginably small early universe entered a state called the Planck era, a period where the conditions were "so extreme that our current physics is completely unable to describe it."[415] Physicist William Stoeger explains that, like the initial singularity, the Planck era represents "a limit falling outside the reliability" of the best current scientific model, and that "a new physics is needed" to understand it.[416] To assert that science will someday be able to adequately describe the extreme conditions of the Planck era or see "before" or "beyond" the initial singularity is not a statement that is grounded in current science but, rather, in a philosophical faith that all things can and will ultimately be understood and that all current physical mysteries are, in principle, comprehensible to the human mind.

412 Brad S. Gregory, "No Room for God: History, Science, Metaphysics, and the Study of Religion," *History and Theory* 47:4 (Dec. 2008), 495–519, 506.

413 Alister E. McGrath, "The Ideological Uses of Evolutionary Biology in Recent Atheist Apologetics," in *Biology and Ideology from Descartes to Dawkins*, ed. Denis R. Alexander and Ronald L. Numbers (Chicago: University of Chicago Press, 2010), 340.

414 For a discussion see Michael Heller "The Initial Singularity and the Creation of the World," in *Ultimate Explanations of the Universe* (London: Springer, 2009).

415 William Stoeger, "God, Physics, and the Big Bang," in *The Cambridge Companion to Science and Religion*, ed. Peter Harrison (Cambridge: Cambridge University Press, 2010), 173.

416 Stoeger, "God, Physics, and the Big Bang," 176.

A second fundamental empirical limit within cosmology concerns how far astronomers can ever hope to see. Given that the universe is expanding and that the speed of light is finite, one can never see further than the distance light has travelled since the big bang. Mathematical physicist John Barrow explains, "Astronomy can only tell us about the structure of the *visible* universe ... we can know nothing of what lies beyond our horizon." Thus, while one may be able confidently to say that the observable universe has certain properties and characteristics, "we can say nothing about the properties of the Universe *as a whole* unless we smuggle in an assumption that the Universe beyond our horizon is the same, or approximately the same, in nature as the visible universe within our horizon."[417] This fundamental limit to human powers of observation—a limit that is imposed by the very nature of light and structure of the cosmos—"prevents us from making any testable statements about the initial structure, or the origin, of the whole Universe."[418] Considering such cases, Barrow consequently concludes, "There are limits to what we can know about the Universe" and "those limits cut across all the major unsolved problems of cosmology."[419]

While cosmology reveals that there are limits to what scientists can know when studying the largest known phenomenon (the whole universe), quantum physics has revealed that there are also limits to what scientists can know when they study the smallest conceivable objects (atoms and their constituent parts). In classical physics, which was the standard view of physics before 1900, it was believed that it was possible simultaneously to know both the position and motion of a given particle with complete accuracy. While the precision of a classical physicist might in practice be limited only by the available technology, there was no reason in principle to expect that better technology would not eventually overcome such limits. According to the standard view of current quantum physics, however, "even with perfect instruments it is impossible to measure the location and velocity of a body simultaneously" with perfect precision. As noted previously, this fundamental limit on the accuracy of measurement is known as Heisenberg's uncertainty principle.[420] "The quantum picture of reality," explains Barrow, "introduces a new form of impossibility into our picture of the world. This impossibility replaces a past belief in unrestricted experimental investigation of Nature which was based upon a

417 John Barrow, *Impossibility: The Limits of Science and the Science of Limits* (Oxford: Oxford University Press, 1998), 160.

418 Barrow, *Impossibility*, 160.

419 Barrow, *Impossibility*, 159.

420 Barrow, *Impossibility*, 22–23.

misconception of what existed to be measured." With quantum physics, says philosopher of science Michael Ruse, "we seem to have reached an outer point of what we can know. This is not a 'science stopper' in the sense of an appeal to miracles ... but it is a recognition that there are areas where we cannot go."[421]

Conclusion

Because non-empirical shaping principles are a key part of what science *is*—as Kuhn and many other historians and philosophers of science have shown—science really does entail faith. Such faith *within* science can take a number of different forms, including ethical values, aesthetic principles, philosophical commitments, metaphysical presuppositions, and theological motivations. Philosopher of science Delvin Ratzsch explains, "Doing science requires use of presuppositions involving criteria for theory construction, theory evaluation, and boundaries of concept legitimacy, plausibility structures, and a host of other matters."[422] Since such factors both precede and inform the practice of science, "science itself cannot provide the rational justification for them."[423] For instance, it is *in principle* not possible for one to "set up a scientific experiment to demonstrate that science or a particular scientific method gives an exhaustive account of reality."[424] There are also no properly scientific means or rules that allow one to scientifically choose between competing research paradigms. Science, then, appears to critically depend upon values and presuppositions that are not the result of scientific discovery or testing. Since they "lie somewhere beyond the borders of science," they cannot be directly evaluated through empirical investigation.[425] Many of the values and presuppositions that the practice science is founded upon come either directly or indirectly from specific religious and theological contexts. Today, scientists can either take these presuppositions for granted as givens (albeit in a rather philosophically arbitrary manner), or seek to provide a deeper philosophical justification for them by appealing to the theological worldview that they emerge from. Alternatively, scientists motivated by non-Judeo-Christian or atheistic social contexts may decide to reject specific presuppositions (such as Ockham's

---

421 Ruse, *Science and Spirituality*, 178. Ruse likewise argues, "We cannot—and never will—solve the body-mind problem," 180.

422 Delvin Ratzsch, "The Nature of Science," in *Science and Religion in Dialogue*, ed. M.Y. Stewart (Oxford: Wiley-Blackwell, 2010), 47.

423 Ratzsch, "The Nature of Science," 47.

424 Mikael Stenmark, *Scientism: Science, Ethics, and Religion* (Aldershot: Ashgate, 2001), 22.

425 Ratzsch, "The Nature of Science," 47.

Razor, the rationality of nature, the intelligibility of nature, ontological realism, or the unity of the laws of nature) and seek to find other guiding presuppositions from non-religious or non-Judeo-Christian contexts. For the present, the vast majority of science is produced under the guiding light of theistically derived philosophical presuppositions. As Davies says, "Science began as an outgrowth of theology, and all scientists, whether atheists or theists accept an essentially theological worldview."[426] Whether or not there can be a different science, however, a science with alternative—non-Judeo-Christian or atheistic—guiding assumptions, and whether or not such a science can thrive as a knowledge producing enterprise, will remain to be seen if and when individual scientists decide to give up the classical philosophical assumptions of science in exchange for a novel set of assumptions.

Works Cited

Aechtner, Thomas. "Galileo Still Goes to Jail: Conflict Model Persistence within Introductory Anthropology Materials." *Zygon*, 50:1 (2015): 209–226.

Aquinas, Thomas. *Basic Writings of St. Thomas Aquinas.* Translated by A.C. Pegis. New York: Random House, 1945.

Aristotle. *Posterior Analytics.* Translated by Richard McKeon. Oxford: Clarendon Press, 1963.

Artigas, Mariano. "Three Levels of Interaction between Science and Philosophy." In *Intelligibility in Science*, edited by C. Dilworth, 123–144. Amsterdam: Rodopi, 1992.

Artigas, Mariano. *The Mind of the Universe: Understanding Science and Religion.* London: Templeton Press, 1999.

Artigas, Mariano and Karl Giberson. *Oracles of Science: Celebrity Scientists versus God and Religion.* Oxford: Oxford University Press, 2007.

Atkins, Peter. "The Limitless Power of Science." In *Nature's Imagination: The Frontiers of Scientific Vision*, edited by John Cornwell, 122–31. Oxford: Oxford University Press, 1995.

Bakay, Louis. *Neurosurgeons of the Past.* Illinois: Charles Thomas, 1987.

Baker, Alan. "Simplicity." In *The Stanford Encyclopedia of Philosophy.* (Fall 2013 Edition), Edward N. Zalta (ed.), URL = <https://plato.stanford.edu/archives/fall2013/entries/simplicity/>.

Barbour, Ian G. *Issues in Science and Religion.* Englewood Cliffs, NJ, Prentice-Hall, 1966.

426 Paul Davies, *Are We Alone?* (New York: Basic Books 1995), 138.

Barbour, Ian G. *Religion in an Age of Science*. San Francisco: Harper & Row, 1990.

Barker, Peter and Bernard R. Goldstein. "Theological Foundations of Kepler's Astronomy." In *Science in Theistic Contexts: Cognitive Dimensions, Osiris 16*, edited by John H. Brooke, M.J. Osler, and J.M. Van der Meer, 88–113. Chicago: University of Chicago Press, 2001.

Barrett, Justin L. *Born Believers: The Science of Children's Religious Belief*. New York: Free Press, 2012.

Barrow, John. *Impossibility: The Limits of Science and the Science of Limits*. Oxford: Oxford University Press, 1998.

Barrow, John D., ed. *Fitness of the Cosmos for Life: Biochemistry and Fine-Tuning*. Cambridge: Cambridge University Press, 2008.

Barth, Karl. *The Knowledge of God and the Service of God according to the Teaching of the Reformation Recalling the Scottish Confession of 1560*. Aberdeen: University of Aberdeen, 1937–1938.

Bennett, Gaymon, Martinez J. Hewlett, Ted Peters and Robert John Russell, eds. *The Evolution of Evil*. Gottingen: Vandenhoeck & Ruprecht, 2008.

Bersanelli, Marco and Mario Gargantini. *Galileo to Gell-Mann: The Wonder that Inspired the Greatest Scientists of all Time*. Philadelphia: Templeton Press, 2009.

Blair, Ann. "Natural Philosophy." In *The Cambridge History of Science: Volume 3, Early Modern Science*, edited by Katharine Park and Lorraine Daston, 365–406. Cambridge: Cambridge University Press, 2003.

Blumenfeld-Kosinski, Renate. *Not of Woman Born: Representations of Caesarean Birth in Medieval and Renaissance Culture*. Ithaca, NY: Cornell University Press, 1990.

Bohr, Niels. *The Philosophical Writings of Niels Bohr, Vol. 4: Causality and Complementarity: Epistemological Lessons of Studies in Atomic Physics, The 1949–50 Gifford Lectures*. Woodbridge, CT: Ox Bow Press, 1999.

Bråkenhielm, Carl Reinhold. "Theology and the Origins of Customized Science." In *The Customization of Science: The Impact of Religious and Political Worldviews on Contemporary Science*, edited by Steve Fuller, Mikael Stenmark and Ulf Zackariasson, 121–138. London: Palgrave Macmillan, 2014.

Brooke, John Hedley. "Preface." In *Science in Theistic Contexts: Cognitive Dimensions, Osiris 16*, edited by John H. Brooke, M.J. Osler, and J.M. Van der Meer, vii–xii. Chicago: University of Chicago Press, 2001.

Brooke, John Hedley. *Science and Religion: Some Historical Perspectives*. Cambridge: Cambridge University Press, 1991.

Brooke, John Hedley, M.J. Osler, and J.M. Van der Meer, eds. *Science in Theistic Contexts: Cognitive Dimensions*. *Osiris* 16. Chicago: University of Chicago Press, 2001.

Brooke, John Hedley, and Geoffrey Cantor. *Reconstructing Nature: The Engagement of Science and Religion*. Edinburgh: T & T Clark, 1998.

Brown, Warren S., Nancey Murphy and H. Newton, eds. *Whatever Happened to the Soul?: Scientific and Theological Portraits of Human Nature.* Minneapolis: Fortress, 1998.

Bultmann, Rudolf. *The Presence of Eternity: History and Eschatology.* New York: Harper, 1957.

Caleon, I.S., G. Lopez Wui, and H.P. Regaya. "Personal Beliefs as Key Drivers in Identifying and Solving Seminal Problems: Lessons from Faraday, Maxwell, Kepler and Newton." *Science Education International* 26:1 (2015): 3–23.

Carnap, Rudolph. *The Logical Structure of the World.* London: Routledge and Kegan Paul, 1967.

Carroll, Sean B. *Endless Forms Most Beautiful: The New Science of Evo Devo.* New York: W.W. Norton & Company, 2006.

Cartwright, Nancy. "Is Natural Science 'Natural' Enough? A Reply to Philip Allport." *Synthese* 94 (1993): 291–301.

Cartwright, Nancy. Quoted in Lydia Jaeger, "Nancy Cartwright's Rejection of the Laws of Nature and the Divine Lawgiver." *Science and Christian Belief* 22:1 (2010): 81–86.

Cigola, Michela. "The Influence of the Society of Jesus on the Spread of European Mechanical Knowledge in China in the XVIth and XVIIth Centuries." In *Explorations in the History of Machines and Mechanisms*, 69–79. Amsterdam: Springer, 2012.

Clayton, Philip. *Mind and Emergence: From Quantum to Consciousness.* New York: Oxford University Press, 2004.

Coakley, Sarah. *Sacrifice Regained: Evolution, Cooperation and God.* Oxford: Oxford University Press, 2017.

Crowther, Kathleen. "The Scientific Revolution." In *The Oxford Handbook of Early Modern European History, 1350–1750: Volume II: Cultures and Power*, edited by Hamish Scott, 56–80. Oxford: Oxford University Press, 2015.

Cutler, Alan H. "Nicolaus Steno and the Problem of Deep Time." In *The Revolution in Geology from the Renaissance to the Enlightenment*, edited by Gary D. Rosenberg, 143–48. Boulder, CO: Geological Society of America, 2009.

Davies, Paul. "The intelligibility of nature." In *Scientific Perspectives on Divine Action: Quantum Cosmology and the Laws of Nature*, edited by Robert J. Russell, Nancey Murphy, and C.J. Isham, 149–164. Berkeley, CA: CTNS and Vatican Observatory Publications, 1999.

Davies, Paul. "Taking Science on Faith." *The New York Times* (November 24th, 2007): 61–62.

Davies, Paul. *Are We Alone?* New York: Basic Books, 1995.

Davis, Edward B. "Christianity and Early Modern Science: The Foster Thesis Reconsidered." In *Evangelicals and Science in Historical Perspective*, edited by

David N. Livingstone, D.G. Hart, and Mark A. Noll, 75–95. Oxford: Oxford University Press, 1999.

Davis, Edward. "Is the Big Bang Theory Irreligious?" *Harrisburg Patriot-News* (25 July 2002), A-13.

Dawkins, Richard. *The Selfish Gene.* Oxford: Oxford University Press, 1989.

Deane-Drummond, Celia. *Re-imaging the Divine Image: Humans and Other Animals.* Kitchener, Ontario: Pandora Press, 2014.

Dennett, Daniel. *Breaking the Spell: Religion as a Natural Phenomenon.* New York: Viking, 2006.

Desmond, Adrian, and James Moore. *Darwin's Sacred Cause: How a Hatred of Slavery Shaped Darwin's Views on Human Evolution.* Boston: Houghton Mifflin Harcourt, 2009.

Dirac, Paul. "The Evolution of the Physicist's Picture of Nature." *Scientific American* 208:5 (1963): 45–53.

"Dissection." Wikipedia https://en.wikipedia.org/wiki/Dissection (December 2016).

Dostoevsky, Fyodor. *A Writer's Diary*, vol. 2: 1877–1881, translated by Kenneth Lantz. Evanston: Northwestern University Press, 1994.

Douglas, Heather. "Rejecting the Ideal of Value-Free Science." In *Value-free Science?: Ideals and Illusions*, edited by Harold Kincaid, John Dupré, and Alison Wylie, 120–139. Oxford: Oxford University Press, 2007.

"Draper's conflict between religion and science." *Catholic World* 21:122 (May 1875): 200–213.

Drake, Stillman. "Kepler and Galileo." In *Essays on Galileo and the History and Philosophy of Science*, edited by Noel M. Swerdlow and Trevor Harvey Levere, 340–351. Toronto: University of Toronto Press, 1999.

Drake, Stillman. "Kepler and Galileo." In *Kepler: Four Hundred Years*, edited by A. Beer and P. Beer, 237–47. New York: Pergamon Press, 1975.

Drake, Stillman. *Essays on Galileo and the History and Philosophy of Science*, edited by Noel M. Swerdlow and Trevor Harvey Levere. Toronto: University of Toronto Press, 1999.

Drake, Stillman. *Galileo: A Very Short Introduction.* New York: Oxford University Press, 2001.

Draper, John William. *History of the Conflict between Religion and Science.* New York: Appleton, 1875.

Dupré, John. "Fact and Value." In *Value-free Science?: Ideals and Illusions*, edited by Harold Kincaid, John Dupré, and Alison Wylie, 27–41. Oxford: Oxford University Press, 2007.

Dupré, John. *Human Nature and the Limits of Science.* Oxford: Oxford University Press, 2001.

Eaglestone, Robert. "Postmodernism and ethics against the metaphysics of comprehension." In *The Cambridge Companion to Postmodernism*, edited by Steven Connor, 182–195. Cambridge: Cambridge University Press, 2004.

Eccles, John. *The Human Mystery.* Berlin: Springer, 1980.

Eccles, John. *The Human Psyche.* New York: Routledge, 1992.

Eddington, Arthur. *The Nature of the Physical World.* New York: Macmillan, 1928.

Efron, Noah J. and Menachem Fisch. "Astronomical Exegesis An Early Modern Jewish Interpretation of the Heavens." In *Science in Theistic Contexts: Cognitive Dimensions, Osiris 16*, edited by John H. Brooke, M.J. Osler, and J.M. Van der Meer, 72–87. Chicago: University of Chicago Press, 2001.

Einstein, Albert. "Physics and Reality." In *Ideas and Opinions*, translated by Sonja Bargmann, 290–322. New York: Bonanza, 1954.

Einstein, Albert. Quoted in Graham Farmelo, *It Must be Beautiful: Great Equations of Modern Science*. London: Granta Books, 2002.

Einstein, Albert. Quoted in Leonard Kollender Nash, *The Nature of the Natural Sciences.* Boston: Little Brown, 1963.

Engler, Gideon. "Aesthetics in Science and in Art." *British Journal of Aesthetics* 30 (1990): 24–34.

Feldhay, Rivka. "Religion." In *The Cambridge History of Science: Volume 3, Early Modern Science*, edited by Katharine Park and Lorraine Daston, 727–755. Cambridge: Cambridge University Press, 2003.

Feldhay, Rivka. *Galileo and the Church: Political Inquisition Or Critical Dialogue?* New York: Cambridge University Press, 1995.

Ferngren, Gary. "Introduction." In *Science and Religion: A Historical Introduction*, edited by Gary Ferngren, ix-xiv. Baltimore: Johns-Hopkins University Press, 2002.

Ferngren, Gary. "Preface." In *The History of Science and Religion in the Western Tradition: An Encyclopedia*, edited by Gary B. Ferngren, xiii-xiv. New York: Garland, 2000.

Ferngren, Gary B., ed. *Science and Religion: A Historical Introduction.* Baltimore, MD: Johns Hopkins University Press, 2017.

Feynman, Richard. Quoted in Marco Bersanelli Mario Gargantini, *Galileo to Gell-Mann: The Wonder that Inspired the Greatest Scientists of all Time.* Philadelphia: Templeton Press, 2009.

Findlen, Paula. "Anatomy theaters, botanical gardens, and natural history collections." In *The Cambridge History of Science: Volume 3, Early Modern Science*, edited by Katharine Park and Lorraine Daston, 272–289. Cambridge: Cambridge University Press, 2003.

Finocchiaro, Maurice A. "The Copernican Revolution and the Galileo Affair." In *The Blackwell Companion to Science and Christianity*, edited by J.B. Stump and A.G. Padgett, 14–25. Chichester: Wiley, 2012.

Finocchiaro, Maurice A. "The Galileo Affair from John Milton to John Paul II: Problems and Prospects." *Science and Education* 8:2 (March 1999): 189–209.

Finocchiaro, Maurice A. "Myth 8: That Galileo Was Imprisoned and Tortured for Advocating Copernicanism." In *Galileo Goes to Jail, and Other Myths about Science and Religion*, edited by Ronald L. Numbers, 68–78. Cambridge, MA: Harvard University Press, 2009.

Finocchiaro, Maurice A. "Science, Religion, and the Historiography of the Galileo Affair: On the Undesirability of Oversimplification." *Osiris, Vol. 16, Science in Theistic Contexts: Cognitive Dimensions* (2001): 114–132.

Finocchiaro, Maurice A. *Defending Copernicus and Galileo: Critical Reasoning in the Two Affairs.* New York: Springer, 2010.

Finocchiaro, Maurice A. *The Essential Galileo.* Indianapolis, IN: Hackett Publishing, 2008.

Finocchiaro, Maurice A. *The Galileo Affair: A Documentary History.* Berkeley, University of California Press, 1989.

Finocchiaro, Maurice A. *The Routledge Guidebook to Galileo's Dialogue.* New York: Routledge, 2013.

Fleming, Donald. *John William Draper and the Religion of Science.* Philadelphia: University of Pennsylvania Press, 1950.

Førland, Tor Egil. "Acts of God? Miracles and Scientific Explanation." *History and Theory* 47 (December 2008): 483–494.

Foster, Michael. "The Christian Doctrine of Creation and the Rise of Modern Natural Science." *Mind* 43 (1934): 446–468.

Foster, Michael. "Greek and Christian Ideas of Nature." *The Free University Quarterly* 6 (1959): 122–127.

Funk, Cary and Becka A. Alper. "Religion and Science." 2015 Pew Research Foundation Poll, http://www.pewinternet.org/2015/10/22/science-and-religion/ (October 22, 2015).

Gates, S. James. "The Workings of Science." (AAAS, December 2016) http://www.scienceforseminaries.org/resource/the-workings-of-science/.

Ghosh, Sanjib Kumar. "Human cadaveric dissection: a historical account from ancient Greece to the modern era." *Anatomy & Cell Biology* 48:3 (September 2015): 153–169.

Giere, Ronald N. "Naturalism." In *A Companion to the Philosophy of Science*, edited by W.H. Newton-Smith, 308–310. Oxford: Blackwell, 2000.

Gingerich, Owen and James MacLachlan. *Nicolaus Copernicus: Making the Earth a Planet.* Oxford: Oxford University Press, 2005.

Glacken, Clarence. *Traces on the Rhodian Shore: Nature and Culture in Western Thought from Ancient Times to the End of the Eighteenth Century.* Berkeley: University of California Press, 1967.

Gould, Stephen Jay quoted in Michael Ruse, *Science and Spirituality: Making Room for Faith in the Age of Science.* Cambridge: Cambridge University Press, 2010.

Grant, Edward. *The Nature of Natural Philosophy in the Late Middle Ages.* Washington, DC: Catholic University of America Press, 2010.

Gregory, Brad S. "No Room for God: History, Science, Metaphysics, and the Study of Religion." *History and Theory* 47:4 (Dec. 2008): 495–519.

Hagen, Edward H. "Controversial Issues in Evolutionary Psychology." In *The Handbook of Evolutionary Psychology*, edited by David M. Buss, 145–174. Hoboken, NJ: Wiley, 2005.

Hannam, James. *God's Philosophers: How the Medieval World Laid the Foundations of Modern Science.* Washington, DC: Regnery Publishing, 2011.

Harrison, Peter. "Introduction." In *Cambridge Companion to Science and Religion*, edited by Peter Harrison, 1–17. Cambridge: Cambridge University Press, 2010.

Harrison, Peter. *The Bible, Protestantism, and the Rise of Natural Science.* Cambridge: Cambridge University Press, 1998.

Harrison, Peter. *The Fall of Man and the Foundations of Science.* Cambridge: Cambridge University Press, 2007.

Harrison, Peter. *The Territories of Science and Religion.* Chicago: University of Chicago Press, 2014.

Haught, John. *God After Darwin: A Theology of Evolution.* Boulder, CO: Westview Press, 2000.

"Healing." In the *New Dictionary of Christian Ethics & Pastoral Theology*, edited by David Atkinson and David Field, 486. Downers Grove, IN: InterVarsity Press, 1995.

Heilbron, John L. *Electricity in the 17th and 18th Centuries: A Study of Early Modern Physics.* Berkeley: University of California Press, 1979.

Heilbron, John L. *Galileo.* Oxford: Oxford University Press, 2010.

Heisenberg, Werner. "Letter to Albert Einstein." In Ian Stewart, *Why Beauty is Truth: A History of Symmetry*, 278. New York: Basic Books, 2008.

Heller, Michael. *Ultimate Explanations of the Universe.* London: Springer, 2009.

Hodgson, Peter. "Presuppositions and Limits of Science." In *The Structure and Development of Science*, edited by G. Radnitzky and G. Andersson, Boston Studies in the Philosophy and History of Science, 133–147. Dordrecht: Reidel, 1979.

Hodgson, Peter. *Theology and Modern Physics.* Burlington, VT: Ashgate, 2005.

Holder, Rodney. "Georges Lemaître and Fred Hoyle: Contrasting Characters in Science and Religion." In *Georges Lemaître: Life, Science and Legacy*, ed. Rodney Holder and Simon Mitton, 39–53. New York: Springer, 2013.

Hooykaas, Riejer. *Fact, Faith and Fiction in the Development of Science: The Gifford Lectures Given in the University of St. Andrews 1976.* Dordrecht; Boston: Kluwer Academic Publishers, 1999.

Horgan, John. "Physics Titan Edward Witten Still Thinks String Theory Is 'On the Right Track.'" *Scientific American* (September 29, 2014) URL: https://blogs.scientificamerican.com/cross-check/physics-titan-still-thinks-string-theory-is-on-the-right-track/.

Hugh of St Victor, *Didascalicon.*

Hugh of St Victor, *De tribus diebus.*

Hunter, Michael, ed. *Robert Boyle Reconsidered.* Cambridge: Cambridge University Press, 1994.

Jaeger, Lydia. "Nancy Cartwright's Rejection of the Laws of Nature and the Divine Lawgiver." *Science and Christian Belief* 22:1 (2010): 81–86.

Jaki, Stanley. *The Road of Science and the Ways to God.* Chicago: University of Chicago Press, 1978.

James, William. *The Varieties of Religious Experience: A Study in Human Nature.* New York: Modern Library, 1902.

Jardine, Nicholas. *The Birth of History and the Philosophy of Science: Kepler's A Defense of Tycho against Ursus with Essays on its Provenance and Significance.* Cambridge: Cambridge University Press, 1984.

Jin, Jennie. "The People's Peking Man: Popular Science and Human Identity in Twentieth-Century China." *PaleoAnthropology* (2010): 72–77.

Kaiser, Christopher. "The Creationist Tradition in the History of Science." *Perspectives on Science and Christian Faith* 45 (June 1993), 80–89.

Kaiser, Christopher. "Early Christian Belief in Creation and the Beliefs Sustaining the Modern Scientific Endeavor." In *The Blackwell Companion to Science and Christianity*, edited by J.B. Stump and Alan G. Padgett, 1–13. Malden, MA: Blackwell, 2012.

Kaku. Michio. Interview on the Leonard Lopate Show, WNYC, January 2, 2004.

Keuzenkamp, Hugo A., Michael McAleer and Arnold Zellner. "The enigma of simplicity." In *Simplicity, Inference and Modelling: Keeping it Sophisticatedly Simple*, edited by Arnold Zellner, Hugo A. Keuzenkamp and Michael McAleer, 1–10. Cambridge: Cambridge University Press, 2004.

Koertge, Noretta, ed. *A House Built on Sand: Exposing Postmodernist Myths about Science.* New York: Oxford University Press, 1998.

Kolb, Rocky. *Blind Watchers of the Sky: The People and Ideas that Shaped Our View of the Universe.* Oxford: Oxford University Press, 1999.

Kragh, Helge. *Cosmology and Controversy: The Historical Development of Two Theories of the Universe.* Princeton: Princeton University Press, 1999.

Kuhn, Thomas. *The Essential Tension: Selected Studies in Scientific Tradition and Change.* Chicago: University of Chicago Press, 1977.

Lindberg, David C. "Galileo, the Church, and the Cosmos." In *When Science and Christianity Meet*, edited by David C. Lindberg and Ronald L. Numbers, 33–60. Chicago: University of Chicago Press, 2003.

Lindberg, David and Ronald Numbers, eds. *God and Nature: Historical Essays on the Encounter between Christianity and Science*. Berkeley: University of California Press, 1986.

Lindberg, David and Ronald Numbers, eds. *When Science and Christianity Meet*. Chicago: University of Chicago Press, 2003.

Livingstone, David. *Darwin's Forgotten Defenders: The Encounter Between Evangelical Theology and Evolutionary Thought*. Grand Rapids: Eerdmans, 1987.

Machamer, Peter. "Galileo Galilei." *The Stanford Encyclopedia of Philosophy* (Winter 2014 Edition), edited by Edward N. Zalta. URL = https://plato.stanford.edu/archives/win2014/entries/galileo/.

MacKay, Donald M. *Behind the Eye*. Oxford: Blackwell, 1991.

Margolis, Howard. "Tycho's system and Galileo's *Dialogue*." *Studies in History and Philosophy of Science* 22:2 (1991): 259–275.

Mavrodi, Alexandra, and George Paraskevas. "Mondino de Luzzi: a luminous figure in the darkness of the Middle Ages." *Croatian Medical Journal* 55:1 (February 2014): 50–53.

McAllister, James W. "Is Beauty a Sign of Truth in Scientific Theories?" *American Scientist* 86 (1998): 174–183.

McAllister, James W. "Truth and beauty in scientific reason." *Synthese* 78 (1989): 25–51.

McGrath, Alister E. "The Ideological Uses of Evolutionary Biology in Recent Atheist Apologetics." In *Biology and Ideology from Descartes to Dawkins*, edited by Denis R. Alexander, and Ronald L. Numbers, 329–351. Chicago: University of Chicago Press, 2010.

McGrath, Alister E. *A Fine-Tuned Universe: The Quest for God in Science and Theology*. Louisville, KY: Westminster John Knox, 2009.

McGrath, Alister E. *A Scientific Theology*, Volume 1, *Nature*. Edinburgh: T & T Clark, 2003.

McGrath, Alister E. *A Scientific Theology*, Volume 2, *Reality*. London: T & T Clark, 2006.

McGrath, Alister E. *The Open Secret: A New Vision for Natural Theology*. Maiden, MA and Oxford: Blackwell-Wiley, 2008.

McMullin, Ernan. "Galileo on Science and Scripture." In *The Cambridge Companion to Galileo*, edited by Peter Machamer, 271–347. Cambridge: Cambridge University Press, 1998.

McMullin, Ernan. *Newton on Matter and Activity*. Notre Dame: University of Notre Dame Press, 1978.

Miller, David M. "The Thirty Years War and the Galileo Affair." *History of Science* 46 (March 2008): 49–74.

Mills, David. *Atheist Universe: The Thinking Person's Answer to Christian Fundamentalism*. Berkeley, CA: Ulysses Press, 2006.

Moltmann, Jürgen. *God in Creation: A New Theology of Creation and the Spirit of God.* Minneapolis: Fortress Press, 1993.

Moore, James R. *The Post-Darwinian Controversies: A Study of the Protestant Struggle to Come to Terms with Darwin in Great Britain and America, 1870–1900.* Cambridge: Cambridge University Press, 1979.

Moritz, Joshua M. "Scientific Revolution." In *The Routledge Companion to Modern Christian Thought*, edited by Chad Meister and James Beilby, 264–276. London; New York: Routledge, 2013.

Moritz, Joshua M. *Human Uniqueness and the Image of God: Bridging Biology and Theology through the Hebrew Concept of Election.* Kitchener, Ontario: Pandora Press, 2018.

Moritz, Joshua M. *Science and Religion: Beyond Warfare and Toward Understanding.* Winona, MN: Anselm Academic, 2016.

Morris, Simon Conway. *Life's Solution: Inevitable Humans in a Lonely Universe.* Cambridge: Cambridge University Press, 2003.

Morris, Simon Conway. *The Runes of Evolution: How the Universe Became Self-Aware.* West Conshohocken: Templeton Press, 2015.

Murphy, Nancey. *Bodies and Souls or Spirited Bodies?* Cambridge: Cambridge University Press, 2006.

Murphy, Nancey, George Ellis, and Timothy O'Connor, eds. *Downward Causation and the Neurobiology of Free Will.* Berlin: Springer, 2009.

Murphy, Nancey, Robert J. Russell, and William Stoeger, eds. *Physics and Cosmology: Scientific Perspectives on the Problem of Natural Evil.* Vatican City State: Vatican Observatory Publications, 2007.

Needham, Joseph. *Human Law and the Laws of Nature in China and the West.* L.T. Hobhouse Memorial Trust Lecture, Cambridge University Press, 1951.

Newton, Isaac. *Principia Mathematica.* London, 1687.

Numbers, Ronald. "Science and Religion." *Osiris*, Volume 1, *Historical Writing on American Science* (1985): 59–80.

Numbers, Ronald L., ed. *Galileo Goes to Jail, and Other Myths about Science and Religion.* Cambridge, MA: Harvard University Press, 2009.

O'Malley, John W. "Introduction to Part III." In *The Jesuits II: Cultures, Sciences, and the Arts, 1540–1773*, edited by John W. O'Malley. Toronto: University of Toronto Press, 2006.

O'Malley, John W., Gauvin Bailey, Steven Harris and Frank Kennedy, eds. *The Jesuits: Cultures, Sciences, and the Arts.* Toronto: University of Toronto Press, 1999.

O'Neill, Ynez Violé. "Innocent III and the Evolution of Anatomy." *Medical History* 20 (1976): 429–433.

Osler, Margaret J. *Divine Will and the Mechanical Philosophy: Gassendi and Descartes on Contingency and Necessity in the Created World.* Cambridge: Cambridge University Press, 1994.

Osler, Margaret J. *Reconfiguring the World: Nature, God, and Human Understanding from the Middle Ages to Early Modern Europe.* Baltimore: Johns Hopkins University Press, 2010.

Osborne, Harold. "Mathematical Beauty and Physical Science." *British Journal of Aesthetics* 24 (1984): 291–300.

Padgett, Alan G. "Science and Theology." In *The Encyclopedia of Christianity*, vol. 4, edited by Erwin Fahlbusch, et al. Grand Rapids: Eerdmans/Brill, 2005.

Pannenberg, Wolfhart. *Toward a Theology of Nature: Essays on Science and Faith.* Translated by Ted Peters. Nashville: Westminster/John Knox Press, 1993.

Park, Katharine. "The Criminal and the Saintly Body: Autopsy and Dissection in Renaissance Italy." *Renaissance Quarterly* 47.1 (1994): 1–33.

Park, Katherine. "Myth 5: That the medieval church prohibited human dissection." In *Galileo Goes to Jail: And Other Myths about Science and Religion*, edited by Ronald L. Numbers, 43–49. Cambridge: Harvard University Press: 2009.

Peacocke, Arthur. *Theology for a Scientific Age: Being and Becoming—Natural, Divine, and Human.* Minneapolis: Fortress, 1993.

Pennock, Robert T. "The Postmodern Sin of Intelligent Design Creationism." *Science & Education* 19:6 (June 2010): 757–778.

Peters, Ted. "Theology and the Natural Sciences." *Modern Theologians: An introduction to Christian theology in the twentieth century*, edited by David F. Ford, 2nd ed. Cambridge, MA: Blackwell, 1997.

Peters, Ted. *God–The World's Future: Systematic Theology for a Postmodern Era.* Minneapolis: Fortress Press, 1992.

Peters, Ted. *Playing God?: Genetic Determinism and Human Freedom*, 2nd ed. New York and London: Routledge, 2003.

Peters, Ted, Robert John Russell and Joshua M. Moritz, eds. *Astrotheology: Science and Theology Meet Extraterrestrial Life.* Eugene, Oregon: Wipf and Stock, 2018.

Polanyi, Michael. *Personal Knowledge: Towards a Post-Critical Philosophy.* Routledge, London, 1962.

Polanyi, Michael. *Personal Knowledge: Towards a Post-Critical Philosophy.* Chicago: University of Chicago Press, 1958.

Polkinghorne, John. *The Faith of a Physicist: Reflections of a Bottom-Up Thinker, The Gifford Lectures, 1993–4.* Princeton, NJ: Princeton University Press, 1994.

Polkinghorne, John. *One World: The Interaction of Science and Theology.* Philadelphia: Templeton Foundation Press, 2007.

Polkinghorne, John. *Science and Creation: The Search for Understanding.* London: SPCK, 1988.

Popper, Karl. *The Logic of Scientific Discovery.* London and Cambridge, MA: Unwin Hyman, 1990.

Principe, Lawrence. "The Warfare Thesis." Science and Religion, recorded lecture, Chantilly, VA: The Teaching Company, 2006.

Principe, Lawrence. *The Scientific Revolution: A Very Short Introduction.* Oxford: Oxford University Press, 2011.

Qiu, J. "How China Is Rewriting the Book on Human Origins." *Nature* 12:535 (July 2016): 218–220.

Prothero, Stephen. *The White Buddhist: The Asian Odyssey of Henry Steel Olcott.* Bloomington: Indiana University Press, 1996.

Ratzsch, Delvin. "The Nature of Science." In *Science and Religion in Dialogue*, edited by M.Y. Stewart. Oxford: Wiley-Blackwell, 2010.

Ravetz, J.R. "The Copernican Revolution." In *The Companion to the History of Modern Science*, edited by G.N. Cantor, J.R.R. Christie, M.J.S. Hodge, and R.C. Olby, 201–216. London: Routledge, 1996.

Reiss, Julian, and Jan Sprenger. "Scientific Objectivity." *The Stanford Encyclopedia of Philosophy* (Summer 2016 Edition), Edward N. Zalta (edited by), URL = <https://plato.stanford.edu/archives/sum2016/entries/scientific-objectivity/>.

Rescher, Nicholas. *Epistemology: An Introduction to the Theory of Knowledge.* Albany, NY: SUNY Press, 2003.

Rescher, Nicholas. *Scientific Realism: A Critical Reappraisal.* Dordrecht: Reidel, 1987.

Rolston III, Holmes. *Genes, Genesis and God: Values and their Origins in Natural and Human History.* Cambridge, UK: Cambridge University Press, 1999.

Rudwick, Martin. *Bursting the Limits of Time: The Reconstruction of Geohistory in the Age of Revolution.* Chicago: University of Chicago Press, 2005.

Rudwick, Martin. *Earth's Deep History: How It Was Discovered and Why It Matters.* Chicago: University of Chicago Press, 2014.

Ruse, Michael. "Atheism and Science." In *The Customization of Science: The Impact of Religious and Political Worldviews on Contemporary Science*, edited by Steve Fuller, Mikael Stenmark and Ulf Zackariasson, 73–88. London: Palgrave Macmillan, 2014.

Ruse, Michael. *Can a Darwinian Be a Christian?: The Relationship Between Science and Religion.* Cambridge: Cambridge University Press, 2001.

Ruse, Michael. *Science and Spirituality: Making Room for Faith in the Age of Science.* Cambridge: Cambridge University Press, 2010.

Russell, Colin. "The Conflict of Science and Religion." In *Science and Religion: A Historical Introduction*, edited by Gary Ferngren, 3–12. Baltimore: Johns-Hopkins University Press, 2002.

Russell, Colin. *Michael Faraday: Physics and Faith.* New York: Oxford University Press, 2000.

Russell, Robert J. *Cosmology from Alpha to Omega: The Creative Mutual Interaction of Theology and Science.* Minneapolis: Fortress, 2008.

Russell, Robert J., Nancey C. Murphy and Arthur R. Peacocke, eds. *Chaos and Complexity: Scientific Perspectives on Divine Action.* Vatican City State; Berkeley, CA: Vatican Observatory Publications; Center for Theology and the Natural Sciences, 1995.

Russell, Robert J., Francisco J. Ayala, and William R. Stoeger, SJ, eds. *Evolutionary Molecular Biology: Scientific Perspectives on Divine Action.* Vatican City State; Berkeley, CA: Vatican Observatory Publications; Center for Theology and the Natural Sciences, 1998.

Russell, Robert J., Nancey Murphy, Theo Meyering and Michael Arbib, eds. *Neuroscience and the Person: Scientific Perspectives on Divine Action.* Vatican City State; Berkeley, CA; Vatican Observatory; Center for Theology and the Natural Sciences, 1999.

Russell, Robert J., Philip Clayton, Kirk Wegter-McNelly, and John Polkinghorne, eds. *Quantum Mechanics: Scientific Perspectives on Divine Action.* Vatican City State; Berkeley, CA: Vatican Observatory Publications; Center for Theology and the Natural Sciences, 2001.

Russell, Robert J., Nancey Murphy, and C.J. Isham, eds. *Scientific Perspectives on Divine Action: Quantum Cosmology and the Laws of Nature.* Vatican City State; Berkeley, CA: Vatican Observatory Publications; Center for Theology and the Natural Sciences, 1999.

Sautman, Barry. "Peking Man and the Politics of Paleoanthropological Nationalism in China." *The Journal of Asian Studies* 60:1 (Feb. 2001): 95–124.

Savage-Smith, Emilie. "Attitudes Toward Dissection in Medieval Islam." *Journal of the History of Medicine and Allied Sciences* 50: 1 (1 January 1995): 67–110.

Schroer, Bert. "String theory, the crisis in particle physics and the ascent of metaphoric arguments." *International Journal of Modern Physics D Int. J. Mod. Phys. D*, 17, 2373 (14 Mar 2006) http://arxiv.org/abs/physics/0603112.

Schwarz, John. "Superstring-A Brief History." In *History of Original Ideas and Basic Discoveries in Particle Physics*, edited by H. Newman and T. Ypsilantis, 695–706. New York: Plenum Press, 1996.

Schweitzer, Albert. *The Problem of Natural Theology and Natural Ethics.* Unpublished.

Shapin, Steven, and Simon Schaffer. *Leviathan and the Air Pump.* Princeton: Princeton University Press, 1985.

Shea, William R., and Mariano Artigas. *Galileo in Rome: The Rise and Fall of a Troublesome Genius.* Oxford: Oxford University Press, 2003.

Simon, Herbert A. "Science seeks parsimony, not simplicity: searching for pattern in phenomena." In *Simplicity, Inference and Modeling: Keeping it Sophisticatedly Simple*, edited by Arnold Zellner, Hugo A. Keuzenkamp and Michael McAleer, 32–72. Cambridge: Cambridge University Press, 2004.

Sober, Elliott. *From a Biological Point of View: Essays in Evolutionary Philosophy.* New York: Cambridge University Press, 1994.

Solomon, Robert C. "Business Ethics." In *A Companion to Ethics*, edited by Peter Singer. Malden, Mass.: Blackwell, 1991.

Southgate, Christopher. *The Groaning of Creation: God, Evolution, and the Problem of Evil.* Louisville, KY: Westminster John Knox, 2008.

Stannard, Russell. *The God Experiment: Can Science Prove the Existence of God?* Mahwah, N.J.: Paulist Press, 2016.

Stenmark, Mikael. "Science and the Limits of Knowledge." In *Clashes of Knowledge: Orthodoxies and Heterodoxies in Science and Religion*, edited by Peter Meusburger, Michael Welker, and Edgar Wunder, 111–120. New York: Springer, 2008.

Stenmark, Mikael. *Scientism: Science, Ethics and Religion*. Aldershot: Ashgate, 2001.

Stoeger, William. "God, Physics, and the Big Bang." In *The Cambridge Companion to Science and Religion*, edited by Peter Harrison, 173–189. Cambridge: Cambridge University Press, 2010.

Susskind, Leonard. "Quark Confinement." In *The Rise of the Standard Model: Particle physics in the 1960s and 1970s*, edited by Lillian Hoddeson, L. Brown, M. Riordan, and M. Dresden, 233–243. Cambridge: Cambridge University Press, 1997.

Swerdlow, Noel M. "Galileo's Discoveries with the Telescope and their Evidence for the Copernican Theory." In *The Cambridge Companion to Galileo*, edited by Peter Machamer, 244–270. Cambridge: Cambridge University Press, 1998.

Swinburne, Richard. *The Evolution of the Soul*. Oxford: Oxford University Press, 1986.

Tillich, Paul. *Systematic Theology*. Chicago: University of Chicago Press, 1951–1963.

Torrance, Thomas F. "The Transfinite Significance of Beauty in Science and Theology." In *L'art, la science et la métaphysique: Études offertes à André Mercier*, edited by Luz García Alonso, Evanghelos Moutsopoulos, and Gerhard Seel, 393–418. Berne: Peter Lang, 1993.

Trigg, Roger. "Realism." In *Encyclopedia of Science and Religion*, edited by J. Wentzel van Huyssteen. New York: Macmillan Reference, 2003.

Trigg, Roger. *Beyond Matter: Why Science Needs Metaphysics*. West Conshohocken: Templeton Press, 2015.

Trigg, Roger. *Rationality and Science: Can Science Explain Everything?* Oxford: Blackwell, 1993.

Van Bendegem, Jean Paul. "Why the Largest Number Imaginable is Still a Finite Number." *Logique et Analyse* 42 (2002): 107–126.

van Huyssteen, J. Wentzel. *Alone in the World?: Human Uniqueness in Science and Theology*. Grand Rapids: Eerdmans, 2006.

van Woudenberg, René. "Christianity and Science." In *The Customization of Science: The Impact of Religious and Political Worldviews on Contemporary Science*, edited by Steve Fuller, Mikael Stenmark and Ulf Zackariasson, 55–72. London: Palgrave Macmillan, 2014.

Wadman, Meredith. "One in Three Scientists Confesses to Having Sinned." *Nature* 435, no. 7043 (June 9, 2005): 718–719.

Wallace, William A. "Galileo's Trial and the Proof of the Earth's Motion." *Catholic Dossier* 1:2 (1996): 7–13.

Watson, J.D. *The Double Helix: A Personal Account of the Discovery of the Structure of DNA*, edited by G.S. Stent. London: Weidenfeld and Nicolson, 1968.

Weinberg, Steven. *Dreams of a Final Theory.* New York: Pantheon Books, 1992.

White, Andrew Dickson. *Popular Science Monthly* Part I: 39 (May 1891).

White, Andrew Dickson. *A History of the Warfare of Science with Theology in Christendom.* Reprint edition. Gloucester, MA: Peter Smith, 1978.

Whitehead, Alfred North. *Process and Reality: An Essay in Cosmology.* New York: Free Press, 1929.

Wigner, Eugene. "The Unreasonable Effectiveness of Mathematics in the Natural Sciences." *Communications in Pure and Applied Mathematics* 13:1 (February 1960): 1–14.

Wilkinson, David. *Science, Religion, and the Search for Extraterrestrial Intelligence.* Oxford: Oxford University Press, 2013.

Wilson, David B. "The Historiography of Science and Religion." In *The History of Science and Religion in the Western Tradition: An Encyclopedia*, edited by Gary B. Ferngren, 3–12. New York: Garland, 2000.

Wilson, David B. "The Historiography of Science and Religion." In *Science and Religion: A Historical Introduction*, edited by Gary Ferngren, 13–29. Baltimore: Johns-Hopkins University Press, 2002.

Wilson, Edward. *Consilience: The Unity of Knowledge.* New York: Knopf, 1999.

Wojcik, Jan. *Robert Boyle and the Limits of Reason.* Cambridge: Cambridge University Press, 1997.

Wykstra, Stephen J. "Religious Beliefs, Metaphysical Beliefs, and Historiography of Science." *Osiris*, Vol. 16, *Science in Theistic Contexts: Cognitive Dimensions* (2001): 29–46.

Yalçinkaya, M. Alper. "Science as an ally of religion: a Muslim appropriation of 'the conflict thesis.'" *British Journal for the History of Science* 44: 2 (2011): 161–181.

Yen, Hsiao-pei. "Evolutionary Asiacentrism, Peking Man, and the Origins of Sinocentric Ethno-Nationalism." *Journal of the History of Biology* 47 (2014): 585–625.

Yen, Hsiao-pei. "From palaeoanthropology in China to Chinese palaeoanthropology: Science, imperialism and nationalism in North China, 1920–1939." *History of Science* 53:1 (2012): 21–56.

www.ingramcontent.com/pod-product-compliance
Lightning Source LLC
Chambersburg PA
CBHW071354230926
51846CB00002B/545
9789004360211